Q
180
.A3
H65
2005

35.00

D1613436

WITHDRAWN

MEDIA CENTER
ELIZABETHTOWN COMMUNITY &
TECHNICAL COLLEGE
600 COLLEGE ST. RD.
ELIZABETHTOWN, KY 42701

VICTORY AND VEXATION
IN SCIENCE

VICTORY AND VEXATION IN SCIENCE

Einstein, Bohr, Heisenberg, and Others

GERALD HOLTON

Harvard University Press
Cambridge, Massachusetts
London, England
2005

Copyright © 2005 by the President and Fellows of Harvard College
All rights reserved
Printed in the United States of America

Library of Congress Cataloging-in-Publication Data

Holton, Gerald James.
Victory and vexation in science :
Einstein, Bohr, Heisenberg, and others / Gerald Holton.
p. cm.
Includes bibliographical references and index.
ISBN 0-674-01519-3 (alk. paper)
1. Science. 2. Research. 3. Scientists—20th century. I. Title.
Q180.A3H65 2005
500—dc22   2004060572

To Thomas and Stephan

Contents

Preface ix

Part I Scientists

1 Einstein's Third Paradise 3
2 The Woman in Einstein's Shadow, and a First Glimpse of Einstein's Mind at Work 16
3 Werner Heisenberg and Albert Einstein 26
4 Bohr, Heisenberg, and What Michael Frayn's *Copenhagen* Tries to Tell Us 36
5 Enrico Fermi and the Miracle of the Two Tables 48
6 B. F. Skinner, P. W. Bridgman, and the "Lost Years" 65
7 I. I. Rabi as Educator and Science Warrior 81

Part II Science in Context

8 Paul Tillich, Albert Einstein, and the Quest for the Ultimate 95
9 Henri Poincaré, Marcel Duchamp, and Innovation in Science and Art 115
10 Perspectives on the Thematic Analysis of Scientific Thought 135

11	The Imperative for Basic Science That Serves National Needs	152
12	The Rise of Postmodernisms and the "End of Science"	162
13	Different Perceptions of "Good Science," and Their Effects on Careers of Women Scientists	181
14	"Only Connect": Bridging the Institutionalized Gaps between the Humanities and Sciences in Teaching	194
	Acknowledgments	211
	Index	213

Preface

Never has the power of scientific research to solve old problems and to uncover new ones been more evident than it is today. Nor have there been more publications about science and scientists added year by year. Yet, there is still widespread ignorance about an aspect of the life of the mind among scientists that is essential for a full understanding of their achievements and limitations: the larger *contexts* within which a specific research result is obtained, above all the context of personal interactions that can lead either to success or disagreement, to victory or vexation.

To clarify this point, let me offer an analogy from the philosophy of science. It is now widely agreed in that field that when a point of scientific theory, prediction, or explanation is being tested experimentally, that inquiry involves not merely the specific point being addressed. Rather, the test extends also to the whole body of scientific beliefs within which the particular question is being raised. In the same sense, any examination of the work or biographical details of a scientist will be more complete if attention is also paid to the larger context within which the narrow question is being investigated. In such a case, the context may take on a variety of forms. For example, the point of view some scientists adopt in their work or in their social commitments may become much clearer if matched up against an account of the opposing points of view held at that time by other scientists. Moreover, bringing to light the conflict of one with another, which results respectively in success or disappointment, makes more evident the basis on which different worldviews operate and are defended.

That is the story behind several of the chapters in this book—for example, the conflicts between Werner Heisenberg on the one hand and Albert Einstein (Chapter 3) and Niels Bohr (Chapter 4) on the other, as well as between P. W. Bridgman and B. F. Skinner (Chapter 6). Even the findings in Chapter 13 of the unhappy divergence between what some scientists regard as "good science" and the effects of that view on their careers qualifies under this heading.

Another kind of context is the effect, for good or ill, of the change of external or internal conditions experienced by scientists in the course of time. Thus, Chapter 1 offers an account of how Einstein's intellectual and emotional "paradise" changed from childhood to maturity and to his late period, and Chapter 5 indicates how the trajectory of Enrico Fermi's scientific work progressed as a function of time and in the changing conditions of work in Italy. Similar attention is paid to the time-dependent context within which ruling ideas about proper education in science have come and gone (Chapter 14); I. I. Rabi's change of primary attention from superb science to education and scientific statesmanship (Chapter 7); and the emergence, in the last few decades, of the urgent need to connect the scientific research agenda to societal needs (Chapter 11).

Among the remaining chapters, I note two with special emphasis. One is Chapter 12, "The Rise of Postmodernisms and the 'End of Science.'" It deals with the historical development and claims of a widespread rebellion against central traditions of Western thought on which science has based its methods and authority. It is, again, the context within which to understand dismissals of scientific findings, not only in branches of academe but more ominously by administrators of government (as in their endorsement of "creationism" vs. evolution or in their disregard of global warming).

The other chapter deserving special mention here is the tenth, a survey of the role of scientists' presuppositions (themata) that, even unconsciously, often set the contexts within which scientific thinking progresses, especially in its nascent phase. I have been studying and writing on that conception for many years, and it is with pleasure that I note it has also helped others to explain puzzles about the advancement of science: why some scientists cling stubbornly to ideas that are being severely challenged; why at any given time, even in the mature phase of a science, there exists no single "paradigm" but rather a spectrum of competing points of view, especially among the best practitioners; and why, on the whole, so much good science has been based, from antiquity to today, on a relatively small number of thematic presuppositions.

The book itself reflects the wide-ranging scope of attention and research that, in my opinion, can illuminate the study of even a fairly limited event in the history of science and related fields.

It remains for me to acknowledge gladly the helpful discussions with many of my colleagues and students, and especially the continued, indispensable help I received from three persons: Joan Laws, my trusted Assistant; Gerhard Sonnert, unfailing Research Associate; and Michael Fisher, Editor in Chief of Harvard University Press, skillful and patient as always.

VICTORY AND VEXATION
IN SCIENCE

I

Scientists

— 1 —

Einstein's Third Paradise

Historians of modern science have good reasons to be grateful to Paul Arthur Schilpp, professor of philosophy and Methodist clergyman but better known as the editor of a series of volumes on "Living Philosophers," including several on scientist-philosophers. His motto was: "The asking of questions about a philosopher's meaning while he is alive." And to his everlasting credit, he persuaded Albert Einstein to do what he had resisted all his years: to sit down to write, in 1946 at age sixty-seven, an extensive autobiography–forty-five pages long in print.

To be sure, Einstein excluded there most of what he called "the merely personal." But on the very first page he shared a memory that will guide us to the main conclusion of this chapter. He wrote that when still very young, he had searched for an escape from the seemingly hopeless and demoralizing chase after one's desires and strivings. That escape offered itself first in religion. Although brought up as the son of "entirely irreligious (Jewish) parents," through the teaching in his Catholic primary school, mixed with private instruction in elements of the Jewish religion, he found within himself a "deep religiosity"—indeed, "the religious paradise of youth."

The accuracy of this memorable experience is documented in other sources, including the biographical account of Einstein's sister, Maja. There she makes a plausible extrapolation: that Einstein's "religious feeling" found expression in later years in his deep interest in and actions to ameliorate the difficulties to which fellow Jews were being subjected, actions ranging from his fights against anti-Semitism to his embrace of a version of Zionism, on condition, as he put it in one of his speeches (April 20, 1935), that it would include a "peaceable and friendly cooperation with the Arab people." As we shall see, Maja's extrapolation of the reach of her brother's early religious feelings might well have gone much further.

The primacy of young Albert's First Paradise came to an abrupt end. As he put it early in his "Autobiographical Notes," through reading popular science books he came to doubt the stories of the Bible. Thus he passed first through what he colorfully described as a "positively fanatic indulgence in free thinking."[1] But then he found new enchantments. At age twelve, he read a little book on Euclidean plane geometry—he called it "holy," a veritable "Wonder." Then, still as a boy, he became entranced by the contemplation of that huge external, extra-personal world of science, which presented itself to him "like a great, eternal riddle." To that study one could devote oneself, finding thereby "inner freedom and security." He believed that choosing the "road to this Paradise," though quite antithetical to the first one, and less alluring, did prove itself trustworthy. Indeed, by age sixteen, he had his father declare him to the authorities as "without confession," and for the rest of his life he tried to dissociate himself from organized religious activities and associations, inventing his own form of religiousness, just as he was creating his own physics.

These two realms appeared to him eventually not as separate as numerous biographers would suggest. On the contrary, my task here is to demonstrate that at the heart of Einstein's mature identity there developed a fusion of his First and his Second Paradise—into a Third Paradise, where the meaning of a life of brilliant scientific activity drew on the remnants of his fervent first feelings of youthful religiosity.

A Passion for Variety

For this purpose, we shall have to make what may seem like an excursus, but one that will in the end throw light on Einstein's overwhelming passion, throughout his scientific and personal life, to bring about the joining of these and other seemingly incommensurate aspects, whether in nature or society. In 1918 he gave a glimpse of it in a speech (*"Prinzipien der Forschung"*) honoring the sixtieth birthday of his friend and col-

---

1. All translations from the original German are this author's, where necessary.

league Max Planck, to whose rather metaphysical conception about the purpose of science Einstein had drifted while moving away from the quite opposite, positivistic one of an early intellectual mentor, Ernst Mach. As Einstein put it in that speech, the search for one "simplified and lucid image of the world" not only was the supreme task for a scientist, but also corresponded to a psychological need: to flee from personal, everyday life, with all its dreary disappointments, and escape into the world of objective perception and thought. Into the formation of such a world picture the scientist could place the "center of gravity of his emotional life [*Gefühlsleben*]." And in a sentence with special significance, he added that persevering on the most difficult scientific problems requires "a state of feeling [*Gefühlszustand*] similar to that of a religious person or a lover."

Throughout Einstein's writings, one can watch him searching for that world picture, for a comprehensive *Weltanschauung,* one yielding a total conception that, as he put it, would include every empirical fact *(Gesamtheit der Erfahrungstatsachen)*—not only of physical science, but also of life. Einstein was of course not alone in this pursuit. The German literature of the late nineteenth and early twentieth centuries contained a seemingly obsessive flood of books and essays on the oneness of the world picture. It included writings by both Ernst Mach and Max Planck and, for good measure, a 1912 general manifesto appealing to scholars in all fields of knowledge to combine their efforts in order to "bring forth a comprehensive *Weltanschauung.*" The thirty-four signatories included Ernst Mach, Sigmund Freud, Ferdinand Tönnies, David Hilbert, Jacques Loeb—and the then still little-known Albert Einstein.

But while for most others this culturally profound longing for unity —already embedded in the philosophical and literary works they all had studied—was mostly the subject of an occasional opportunity for exhortation (nothing came of the manifesto), for Einstein it was different, a constant preoccupation responding to a persistent, deeply felt intellectual and psychological need.

This fact can be most simply illustrated in his scientific writings. As a first example, I turn to one of my favorite manuscripts in Einstein's archive. It is a lengthy manuscript in his handwriting, of around 1920, titled, in translation, "Fundamental Ideas and Methods of Relativity." It contains the passage in which Einstein reveals what in his words was "the happiest thought of my life"—*der glücklichste Gedanke meines Lebens*. This refers to a thought experiment that came to him in 1907: nothing less than the definition of the equivalence principle, later developed in his general relativity theory. It occurred to Einstein—thinking first of all in visual terms, as was usual for him—that if a man were falling from the roof of his house and tried to let anything drop, it would only move alongside him, thus indicating the equivalence of acceleration and gravity. In Einstein's words, "the acceleration of free fall with respect to the material is therefore a mighty argument that the postulate of relativity is to be extended to coordinate systems that move non-uniformly relative to one another."

For the present purpose I want to draw attention to another passage in that manuscript. The essay begins in a largely impersonal, pedagogic tone, similar to that of his first popular book on relativity, published in 1917. But in a surprising way, in the section titled "General Relativity Theory," Einstein suddenly switches to a personal account. He reports that in the construction of the special theory, the "thought concerning the Faraday [experiment] on electromagnetic induction played for me a leading role." He then describes that old experiment, in words similar to the first paragraph of his 1905 relativity paper, concentrating on the well-known fact, discovered by Faraday in 1831, that the induced current is the same whether it is the coil or the magnet that is in motion relative to the other, whereas the "theoretical interpretation of the phenomenon in these two cases is quite different." While other physicists, for many decades, had been quite satisfied with that difference, here Einstein revealed a central preoccupation at the depth of his soul: "The thought that one is dealing here with two fundamentally different cases was for me unbearable [*war mir unerträglich*]. The difference between these two cases could not be a real difference.... The phenomenon of the electromagnetic induction forced me to postulate the (special) relativity principle."

Let us step back for a moment to contemplate that word *unbearable*. It is reinforced by a passage in Einstein's "Autobiographical Notes": "By and by I despaired [*verzweifelte ich*] of discovering the true laws by means of constructive efforts based on known facts. The longer and the more despairingly I tried, the more I came to the conviction that only the discovery of a universal formal principle could lead us to assured results." He might have added that the same postulational method had already been pioneered in their main works by two of his heroes, Euclid and Newton.

Other physicists, for example Bohr and Heisenberg, also reported that at times they were brought to despair in their research. Still other scientists were evidently even brought to suicide by such disappointment. For researchers fiercely engaged at the very frontier, the psychological stakes can be enormous. Einstein was able to resolve his discomfort by turning, as he did in his 1905 relativity paper, to the *postulation* of two formal principles (the principle of relativity throughout physics, and the constancy of the velocity of light in vacuo), and adopting such postulation as one of his tools of thought.

Einstein also had a second method to bridge the unbearable differences in a theory: *generalizing it,* so that the apparently differently grounded phenomena are revealed to be coming from the same base. We know from a letter to Max von Laue of January 17, 1953, found in the archive, that his early concern with the physics of fluctuation phenomena was the common root of his three great papers of 1905, on such different topics as the quantum property of light, Brownian movement, and relativity. But even earlier, in a letter of April 14, 1901, to his school friend, Marcel Grossmann, Einstein had revealed his generalizing approach to physics while working on his very first published paper, on capillarity. There he tried to bring together in one theory the opposing behaviors of bodies: moving upward when a liquid is in a capillary tube, but downward when the liquid is released freely. In that letter, he spelled out his interpenetrating emotional and scientific needs in one sentence: "It is a wonderful feeling [*ein herrliches Gefühl*] to recognize the unity of a complex of appearances which, to direct sense experiences, appear to be quite separate things."

The postulation of universal formal principles and the discovery among phenomena of a unity, of *Einheitlichkeit*, through the *generalization* of the basic theory—those were two of his favorite weapons,[2] as his letters and manuscripts show. Writing to Willem de Sitter on November

---

2. A third was his use of freely adopted (non-Kantian) categories, or thematic presuppositions. The prominent ones include unity or unification; logical parsimony and necessity; symmetry; simplicity; causality; completeness of explanation; continuum; and, of course, constancy and invariance.

4, 1916, he confessed: "I am driven by my need to generalize [*mein Verallgemeinerungsbedürfnis*]." That need, that compulsion, was also deeply entrenched in German culture and resonated with, and supported, Einstein's approach. Let me just note in passing that while he was still a student at the Polytechnic Institute in Zurich, in order to get his certificate to be a high school science teacher, Einstein took optional courses on Immanuel Kant and Goethe, whose central works he had studied since his teenage years.

That *Verallgemeinerungsbedürfnis* was clearly a driving force behind Einstein's career trajectory. Thus he generalized from old experimental results, like Faraday's, to arrive at special relativity, in which he unified space and time, electric and magnetic forces, energy and mass, and so resolved the whole long dispute among scientists between adherence to a mechanistic versus an electromagnetic world picture. Then he generalized the special theory to produce what he first significantly called (in an article of 1913) not the *general* but the *generalized relativity theory*. Paul Ehrenfest wrote him in puzzlement: "How far will this *Verallgemeinerung* go on?" And, finally, Einstein throws himself into the attempt of a grand unification of quantum physics and of gravity: a unified field theory. It is an example of an intense and perhaps unique, lifelong, tenacious dedication, despite Einstein's failure at the very end. Nevertheless, as a program, it set the stage for the ambition of some of today's best scientists, who have taken over that search for the Holy Grail of physics—a theory of everything.

So much for trying to get a glimpse of the mind of Einstein as scientist. But at this point, for anyone who has studied this man's work and life in detail, a new thought urges itself forward. As in his science, Einstein also *lived* under the compulsion to unify—in his politics, in his social ideals, even in his everyday behavior. He abhorred all nationalisms and called himself, even while in Berlin during World War I, a European; later he supported the One World movement, dreamed of a unified supernational form of government, helped to initiate the international Pugwash movement of scientists during the Cold War, and was as ready to befriend visiting high school students as the Queen of the Belgians. His instinctive penchant for democracy and dislike of hierarchy and class differences must have cost him dearly in the early days, as when he addressed his chief professor at the Swiss Polytechnic Institute (on whose recommendation his entrance to any academic career would depend) not by any title but simply as "Herr Weber." And at the other end of the spectrum, in Einstein's essay on ethics, he cited Moses, Jesus, and Buddha as equally valid prophets.

No boundaries, no barriers; none in life, as there are none in nature. Einstein's life and his work were so mutually resonant that we recognize both to be carried on together in the service of one grand project—the fusion into one coherency.

"Consecrated Science"

There were also no boundaries or barriers between Einstein's scientific and religious feelings. After passing from the youthful first, religious paradise into his second, immensely productive scientific one, he found in his middle years a fusion of those two motivations—his Third Paradise.

We had a hint of this development in his speech in 1918, where he observed the parallel states of feeling of the scientist and of the "religious person." Other hints come from the countless, well-known quotations in which Einstein referred to God—doing it so often that Niels Bohr had to chide him. Karl Popper remarked that in conversations with Einstein, "I learned nothing . . . he tended to express things in theological terms, and this was often the only way to argue with him. I found it finally quite uninteresting."

But two other reports may point to the more profound layer of Einstein's deepest convictions. One is his remark to one of his assistants, Ernst Straus: "What really interests me is whether God had any choice in the creation of the world." The second is Einstein's reply to a curious telegram.

In 1929, Boston's Cardinal O'Connell branded Einstein's theory of relativity as "befogged speculation producing universal doubt about God and His Creation," and as implying "the ghastly apparition of atheism." In alarm, New York's Rabbi Herbert S. Goldstein asked Einstein by telegram: "Do you believe in God? Stop. Answer paid 50 words." In his response, for which Einstein needed but twenty-five words (in German), he stated his beliefs succinctly: "I believe in Spinoza's God, Who reveals Himself in the lawful harmony of the world, not in a God Who concerns Himself with the fate and the doings of mankind." The rabbi cited this as evidence that Einstein was not an atheist and further declared that "Einstein's theory, if carried to its logical conclusion, would bring to mankind a scientific formula for monotheism." Einstein wisely remained silent on that point.

The good rabbi might have had in mind the writings of the Religion of Science movement, which had flourished in Germany under the distinguished auspices of Ernst Haeckel, Wilhelm Ostwald, and their circle (the *Monistenbund*), and also in America, chiefly in Paul Carus's books and journals, such as *The Monist* and *The Open Court*, which carried the words "Devoted to the Religion of Science" on its masthead.

If Einstein had read Carus's book, *The Religion of Science* (1893), he may have agreed with one sentence in it: "Scientific truth is not profane, it is sacred." Indeed, the charismatic view of science in the lives of some scientists has been the subject of scholarly study, for example in Joseph Ben-David's *Scientific Growth* (1991) and earlier in Robert K. Merton's magisterial book of 1938, *Science, Technology and Society in Seventeenth-Century England*. In the section entitled "The Integration of Religion and Science," Merton noted that among the scientists he studied, "the religious ethic, considered as a social force, so consecrated science as to make it a highly respected and laudable focus of attention." The social scientist Bernard H. Gustin elaborated on this perception, writing that science at the highest level is charismatic because a person "is thought to come into contact with what is essential in the universe." I believe this is precisely why so many who knew little about Einstein's scientific writing flocked to catch a glimpse of him and to this day feel somehow uplifted by contemplating his iconic image.

Starting in the late 1920s, Einstein became more and more serious about clarifying the relationship between his transcendental and his scientific impulses. He wrote several essays on religiosity; five of them, composed between 1930 and the early 1950s, are reproduced in his book *Ideas and Opinions*. In those chapters we can watch the result of a struggle that had its origins in his school years, as he developed, or rather invented, a religion that offered a union with science.

In the evolution of religion, he remarked, there were three developmental stages. At the first, "with primitive man it is above all fear that evokes religious notions. This 'religion of fear' . . . is in an important degree stabilized by the formation of a special priestly caste" that colludes with secular authority to take advantage of it for its own interest. The next step, Einstein held—"admirably illustrated in the Jewish scriptures"—was a moral religion embodying the ethical imperative, "a development [that] continued in the New Testament." Yet it had a fatal flaw: "the anthropomorphic character of the concept of God," easy to grasp by "underdeveloped minds" of the masses, but freeing them of responsibility.

This flaw disappears at Einstein's third, mature stage of religion, toward which he believed mankind is now reaching and which the great spirits (he names Democritus, St. Francis of Assisi, and Spinoza) had already attained–namely, the "cosmic religious feeling" that sheds all anthropomorphic elements. In describing the driving motivation toward that final, highest stage, Einstein uses the same ideas, even some of the same phrases, with which he had celebrated first his religious and then his scientific paradise: "The individual feels the futility of human desires, and aims at the sublimity and marvelous order which reveal themselves both in nature and in the world of thought." "Individual existence impresses him as a sort of prison, and he wants to experience the universe as a single, significant whole." Of course! Here as always, there has to be the intoxicating experience of unification. And so Einstein goes on, "I maintain that the cosmic religious feeling is the strongest and noblest motive for scientific research. . . . A contemporary has said not unjustly that in this materialistic age of ours the serious scientific workers are the only profoundly religious people."

In another one of his essays on religion, Einstein points to a plausible source for his specific formulations: "Those individuals to whom we owe the great creative achievements of science were all of them imbued with a truly religious conviction that this universe of ours is something perfect and susceptible through the rational striving for knowledge. If this conviction had not been a strongly emotional one, and if those searching for knowledge had not been inspired by Spinoza's *amor dei intellectualis*, they would hardly have been capable of that untiring devotion which alone enables man to attain his greatest achievements."

Einstein and the "God of Spinoza"

I believe we can guess at the first time Einstein read Baruch Spinoza's *Ethics (Ethica Ordinae Geometrico Demonstrata)*, a system constructed on the Euclidean model of deductions from propositions. Soon after getting his first real job at the patent office, Einstein joined with two other friends to form a discussion circle, meeting once or twice a week in what they called, with gallows humor, *Akademie Olympia*. We know the list of books they read and discussed. High among them, reportedly at Einstein's suggestion, was Spinoza's *Ethics*, which he read afterwards several times more. When his sister Maja joined him in Princeton in later life and was confined to bed by an illness, he thought that reading to her a good book would help; he chose Spinoza's *Ethics* for that purpose.

By that time Spinoza's work and life had long been important to Einstein. He had written an introduction to a biography of Spinoza (by his son-in-law, Rudolf Kayser, 1946); he had contributed to the *Spinoza Dictionary* (1951); he had referred to Spinoza in many of his letters; and he even had composed a poem in his honor. He admired Spinoza for his independence of mind, his deterministic philosophical outlook, his skepticism about organized religion and orthodoxy—which had resulted in Spinoza's excommunication from his synagogue in 1656—and even for his ascetic preference, which compelled him to remain in poverty and solitude, to live in a sort of spiritual ecstasy, instead of accepting a professorship at the University of Heidelberg. Originally neglected, Spinoza's *Ethics*, published only posthumously, eventually influenced others profoundly, including Friedrich Schlegel, Friedrich Schleiermacher, Goethe (who called him "our common saint"), Albert Schweitzer, and Romain Rolland (who, on reading *Ethics*, confessed, "I deciphered not what he said, but what he meant to say").

For Spinoza, God and nature were one *(deus sive natura)*. True religion was based not on dogma but on a feeling for the rationality and the unity underlying all finite and temporal things, on a feeling of wonder and awe that *generates* the idea of God, but a God who lacks any anthropomorphic conception. As Spinoza wrote in Proposition 15 in *Ethics*, he opposed assigning to God "body and soul and being subject to passions." Hence, "God is incorporeal"—as had been said by others, from

Maimonides on, to whom God was knowable indirectly through His creation, through nature. In other pages of *Ethics,* Einstein could read Spinoza's opposition to the idea of cosmic purpose, and that he favored the primacy of the law of cause and effect—an all-pervasive determinism that governs nature and life—over the notion of "playing at dice," in Einstein's famous remark. And as if he were merely paraphrasing Spinoza, Einstein wrote in 1929 that the perception in the universe of "profound reason and beauty constitute true religiosity; in this sense, and in this sense alone, I am a deeply religious man."

Science as Devotion

Much has been written about the response of Einstein's contemporaries to his Spinozistic "cosmic" religion. For example, the physicist Arnold Sommerfeld recorded in the Schilpp volume that he often felt "that Einstein stands in a particularly intimate relation to the God of Spinoza." But what finally interests us here is to what degree Einstein, having reached his Third Paradise, in which the yearnings for science and religion are joined, may even have found in his own research in physics fruitful ideas emerging from that union. In fact there are at least some tantalizing parallels between passages in Spinoza's *Ethics* and Einstein's publications in cosmology—parallels that the physicist and philosopher Max Jammer, in his book *Einstein and Religion* (1999), considers as amounting to intimate connections. For example, in Part I of *Ethics* ("Concerning God"), Proposition 29 begins: "In nature there is nothing contingent, but all things are determined from the necessity of the divine nature to exist and act in a certain manner." Here is at least a discernible overlap with Einstein's tenacious devotion to determinism and strict causality at the fundamental level, despite all the proofs from quantum mechanics of the *usefulness* of allowing probabilism, at least in the subatomic realm.

There are other such parallels throughout. But what is considered by some as the most telling relationship between Spinoza's propositions and Einstein's physics comes from passages such as Corollary 2 of Proposition 20: "It follows that God is immutable or, which is the same thing, all His attributes are immutable." In a letter of September 3, 1915, to Elsa (his cousin and later his wife), Einstein, having read Spinoza's

Ethics again, wrote, "I think the *Ethics* will have a permanent effect on me."

Two years later, when he expanded his general relativity to include "cosmological considerations," Einstein found to his dismay that his system of equations did "not allow the hypothesis of a spatially closed-ness of the world [*räumliche Geschlossenkeit*]." How did Einstein cure this flaw? By doing something he had done very rarely: making an ad hoc addition, purely for convenience: "We can add, on the left side of the field equation an—for the time being—unknown universal constant,—λ." In fact, it seems that not much harm is done thereby. The constant does not change the covariance; the equation still corresponds with the observation of motions in the solar system ("as long as λ is small"), and so forth. Moreover, the proposed new universal constant λ also determines the average density of the universe with which it can remain in equilibrium and provides the radius and volume of a presumed spherical universe.

Altogether a beautiful, immutable universe—one an immutable God could be identified with. In 1922, however, Alexander Friedmann showed that the equations of general relativity did allow expansion or contraction. And in 1929, Edwin Hubble found by astronomical observations that the universe does expand. Thus Einstein—at least according to the physicist George Gamow—admitted that "inserting λ was the biggest blunder of my life."

Max Jammer and the physicist John Wheeler, both of whom knew Einstein, traced his unusual ad hoc insertion of λ, nailing down that "spatially closed-ness of the world," to a relationship of Einstein's thoughts and Spinoza's propositions. They also pointed to another possible reason for it: In Spinoza's writings, one finds the concept that God would not have made an empty world. But in an expanding universe, in the infinity of time, the density of matter would be diluted to zero in the limit. Space itself would disappear, since, as Einstein put it in 1952, "On the basis of the general theory of relativity . . . space as opposed to 'what fills space' . . . had no separate existence."

Even if all of these suggestive indications of an intellectual, emotional, and perhaps even spiritual resonance between Einstein's and Spinoza's writings were left entirely aside, there still remains Einstein's declaration of his "cosmic religion." That was the end point of his own troublesome pilgrimage in religiosity—from his early vision of his First Paradise, through his disillusionments, to his dedication to find fundamental unity within natural science, and at last to his recognition of science as the devotion, in his words, of "a deeply religious unbeliever"—his final embrace of seeming incommensurables in his Third Paradise.

— 2 —

The Woman in Einstein's Shadow, and a First Glimpse of Einstein's Mind at Work

Research in the history of science tends to pay little attention to the "outsider," whose role usually remains in the shadows while the spotlight is on the "insider." I shall draw attention to a woman to whom every historian of modern physics is indebted, but whose role is now known in any detail only to a mere handful of specialists—a woman who for twenty-seven years spent more time face-to-face with Albert Einstein than perhaps any other person: Helen Dukas, the self-effacing but ever loyal and extraordinarily effective secretary and helpmate of Einstein.

Helen Dukas

From her first day of employment in 1928 to Einstein's death in 1955, and in important ways for many years afterward, Helen Dukas was the person who read and typed Einstein's correspondence, translated with great style Einstein's letters into English, and saw to it that the vast correspondence and manuscripts were saved (in the face of Einstein's own typical disinterest and neglect of such matters). Without her scrupulous collector's passion and devotion to Einstein, we simply would now have only a mere fraction of the collected papers of Einstein that have already sparked so much important scholarship. For many decades she was also a member of the household, and therefore saw at first hand both the bright and dark sides of the life of the Einstein families. On Einstein's death, she became a trustee of his estate, in accord with his last will.

I first met Helen Dukas on August 13, 1959. I had traveled to the

Institute for Advanced Study in Princeton with a recommendation from my colleague, the physicist-philosopher and Einstein's friend and biographer, Philipp Frank, in the hope of consulting some of the documents in the Einstein estate while preparing a paper for a conference. After Einstein's death, she had been relegated to the large, room-sized vault in the basement of Fuld Hall at the Institute. That's where I found her, the whole scene illuminated only by her rather insufficient desk lamp. She was sitting at the desk, bent over some papers; a long row of large file cabinets loomed in the darkness beyond. I could not help but think of Juliet in the crypt, after the death of Romeo.

I asked whether I might look at some of Einstein's documents. Perhaps because of Frank's recommendation, she became very helpful, leading me to the file cabinets, each crammed full of folders of Einstein's correspondence and manuscripts, assembled in an order which only she could have puzzled out. At any rate, the visit went well for my purpose, and indeed perhaps also for hers, because a few days later she wrote me, "It is a real satisfaction to me to be able to help you, and I am looking forward to it eagerly."

Since Helen (as she eventually offered I could call her) is a key both to my story and indeed to the existence of the whole Collected Papers of Einstein, a few more words of description are essential.

She was born on October 17, 1896, in Freiburg im Breisgau, the fourth of seven children. According to a memorial essay by Abraham Pais, she had had to interrupt her Lyceum education at age fifteen, after her mother died, to take charge of running the household and bring up the younger children. Later she became governess in the home of Raphael Straus in Munich, one of whose new nephews was Ernst Straus. Let me give you a taste of this remarkable woman's wit and tough realism. As it happened, in the 1940s Ernst Straus came to the Institute, to be one of Einstein's assistants. When Ernst introduced himself to Helen, she said, "Of course I already know you well: I was present at your circumcision." Or again: In 1965, in one of her letters to me (and we had a correspondence of well over a hundred letters altogether), she said the Russian historian-philosopher Kuznetsov "has sent me the English translation of his new Einstein biography. With my letter of thanks, I enclosed two pages of corrections; *and* I have found since some more." To the President of the Israel Academy of Sciences and Humanities she wrote in 1979, "I looked a little into the catalogue [of an exhibit he had sent]. The Zionist Congress of 1929 took place in Zurich, not München. *I* was there!"[1]

After Einstein's death, Helen attended to his stepdaughter, Margot, in their home at 112 Mercer Street, even while keeping up with the continuing correspondence and inquiries. She was also trying to find new documents, retyping old, fading, or handwritten ones, particularly those in Gothic script. Her sharp memory and her utter devotion and reliability became quickly obvious.

But I am getting ahead of the story. For me, standing in that vault for the first time, looking over this unsuspected, enormous, chaotic treasure of Einstein's papers, I felt like Ali Baba in "The Tales of the Arabian Nights," when he called out "Open Sesame." But even during that first of more than fifty visits, I saw two essential needs. In the absence of serious help, Helen had been trying to type out a catalog of the papers, correspondence, and manuscripts. I came to feel that at the very least, for the sake of the profession of historians of science, one must somehow capture her experience, her memory of the events and correspondence in which she had been involved. No less important than the profession's needs were Helen's own needs. She certainly seemed to have been rather lonely after Einstein's death. As far as I could guess, she was without an office or salary.

Transforming a Collection into an Archive

In supporting my application for funds to help out, John Wheeler put the matter clearly in a letter to Robert Morison of the Rockefeller Foundation. He recommended that support be given for a serious project, namely to put the huge heap of correspondence in good order for use by scholars. Only a few had dared to use it so far. Wheeler wrote, "the great mass of the material is unorganized. Miss Helen Dukas works at this only in a limited way and without assistance or guidance by anyone trained in the history of science. An enormous task requires doing it,

---

1. Her hand in correcting a mistake in Einstein's published "Autobiographical Notes" is discussed in Seiya Abiko, "On Einstein's Distrust of the Electromagnetic Theory: The Origin of the Lightvelocity Postulate," *Historical Studies in the Physical Sciences*, 33, no. 2 (March 2003): 193–215.

and it goes ahead only at a niggling pace." The financial support I looked for (and which was granted) was also needed to microfilm at least the scientific part of the collection. To quote here from one of Helen's letters to me of those days: "The work you have in mind for me fascinates me, but also fills me with apprehension." And in another letter, "I have been hoping for something like this to turn up."

During the many visits, I also had to plant in Helen's mind the idea of eventually allowing publication of the Einstein papers, so as to provide scientists, historians of science, and philosophers of science with the necessary material for their work. Moreover, she had to be made fully aware of the historical value of the riches all around her and persuaded to bring into the vault what she called the "personal stuff." That referred to the more personal letters, which she fought to protect from disclosure but which were in my view needed to supplement the "scientific correspondence." If she had a fault, it was, as Freeman Dyson put it in a memorial essay, that she fiercely guarded the presentation of Einstein as a person "without violent feelings and tragic mistakes," rather than allowing all to see him as "a complete and fully rounded human being."

At any rate, the effort to get more of the "personal stuff" into the available archive succeeded by September 1968, when Helen had a number of file cabinets brought from the house on Mercer Street into the vault, to be included and catalogued. She also made available documents that she had kept in a safe within the vault; it contained, for example, correspondence with Freud, Roosevelt, Romain Rolland, and Elsa's letters. So eventually there were orderly, catalogued folders on Gandhi, Paul Valéry, Bertrand Russell, Chaim Weizmann, the Queen of the Belgians, Tagore, Schweitzer, Thomas Mann, Bernard Shaw, as well as the light-hearted verses of Einstein—all those joining the files of Schrödinger, Pauli, Curie, Lorentz, Bohr, Born, Ehrenfest, Infeld, Hilbert, Bose, de Broglie, Bohm, Debye, Eddington, and so on, to Meitner, Minkowski, and so forth to Wentzel, Wien, and Zeeman. By 1973, 130 such file folders were catalogued, some very bulky, with Ehrenfest's having no less than 165 items. From about 1976 on, the strong editorial staff for the Princeton University Press project greatly expanded what we had started. By the time John Stachel finished in January 1980, he had 42,000 items in his big index, which, as he wrote me, had been based on what he called our "little index."

The foundation money I had raised (and which Harvard administered) was primarily to provide Helen with a salary for her work (which, to her amusement, made her my research assistant). It also gave her the companionship of physics graduate students at Princeton, selected by John Wheeler and myself. These students were hired to come for a few hours or days per week; they did excellent work in helping to prepare a catalogue-index for each item in each file, and they also brightened Helen's life. To help with the work at hand, I made periodic visits to the Institute myself, at least monthly, sometimes weekly, starting in the early 1960s, and staying for longer periods as Member of the Institute in 1964 and as its Visitor in 1971.

Let me confess that I came to respect and even love Helen—much like one of my favorite aunts—and I think a little of such feelings might have been true for her too. We trusted each other fully. After the first few years, whenever she was occasionally ill, she would permit me to work in the vault on my own or to supervise the students, having given me the code for opening the vault and the keys for the files and the safe within. During that whole decade of visits and collaboration and correspondence, I recall not a single time when we were out of sorts with each other. In 1964, she had the idea of giving me a present. It was precious indeed—the set of reprints of Einstein's published papers that had been kept near his desk, and on a few of which he had made corrections and additions. (These are noted in the published *Collected Papers*.) The set was bound in several volumes, and on the first page Helen had written a dedication to me: "To my helper . . ."

Labor's Early Fruits

Of course, all that work helped also in my own research, far beyond my initial hopes. Thanks even to my earliest visits, I was able, in two papers written in 1959, to include (with Helen's permission) key material that had been unpublished and, as it were, was waiting for me in the vault.[2] Throughout my later writings on Einstein, there are many references to

---

2. For example, I used two crucial, previously unpublished Einstein letters in my lectures and writings from 1959 on (see pages 179 and 193 in the 1973 edition of *Thematic Origins*, delivered, respectively, in December 1959 and August 1960).

permissions granted by the Trustees, Helen, and Otto Nathan, allowing me to publish materials from the archive as part of my own work. I have little doubt that without that rather accidental introduction to the papers of Einstein and his circle, my professional life would have been very different.

One day it struck me that my mental preparation for research on Einstein's work had actually begun much earlier. When I was about fifteen years old, my father and I went on habitual searches through bookstores when he had time, and on one such occasion my eyes fell on a battered copy of Einstein's first book on relativity (1917), the one he called *"gemeinverständlich"* (literally, "generally comprehensible"). Here was something I had been waiting for. Undeterred by the previous owner's inscription ("That Jew Einstein stole it all from Lorentz"—after all, this was in Vienna), I tried my best to understand that enchanted part of science which had not yet entered our curriculum. In fact, it was one of the only two books I could take along when I left Vienna about a year later.

The intellectual fruits of work at the Princeton Institute came quickly. One reason for my initial trip to Fuld Hall's vault was that—while there existed many biographies of various qualities, and even Einstein's own extraordinary "Autobiographical Notes" in the Schilpp volume of 1949 —virtually nothing had been done on Einstein by historians of science, a lot of whom seemed still occupied with the never-ending task of glimpsing the minds of scientists of earlier times, such as Galileo and Newton. Someone had to take "the first step forward," to start seriously on Einstein.[3]

In fact, at the time, some works passing for historical analysis were downright misleading. One example is the famous case of Sir Edmund Whittaker, a distinguished physicist known especially for his contributions to classical mechanics. His book, *A History of the Theories of Ether and Electricity,* was first published in 1910; a second edition, with the same title, appeared in the 1950s. It was in many respects a *tour de force.* But in that second edition, completed in 1953, Whittaker revealed his strange view of the origins of Einstein's special theory of relativity. Ein-

---

3. See the discussion on this point by T. Hirosige, "The Ether Problem, the Mechanistic Worldview, and the Origins of the Theory of Relativity," *Historical Studies in the Physical Sciences,* 7 (1976): 3–82, on p. 5.

stein's 1905 paper, Whittaker wrote, was one that "set forth the relativity theory of Poincaré and Lorentz with some amplifications, and which attracted much attention." Moreover, two years later, now writing the necrology for Einstein in the *Biographical Memoirs* of the Fellows of the Royal Society (London, 1955), he had not changed his evaluation concerning Einstein's paper. He concluded with the remarkable sentence: "Einstein [in 1905] adopted Poincaré's name for it, as a new basis for physics, and showed that the group of Lorentz transformations provided a new analysis connecting the physics of bodies in motion relative to each other" (p. 42). In one of my first papers on the history of relativity theory,[4] written after my earliest visits to the Archive, I dedicated several pages to the analysis of Whittaker's perception of the much more complex story.

A second, more well known sort of perspective on the origins of Einstein's theory was, in those years, the generally shared idea that the Michelson-Morley experiment was the "crucial" basis on which he had built his theory (rather than being small and indirect, one of the many negative ether experiments of which young Einstein undoubtedly had read and which made Einstein's work palatable in retrospect for many other physicists). Einstein repeatedly denied the allegation of any crucial influence, but the false story fulfilled nicely the needs of pedagogues to make Einstein's counter-intuitive work plausible to students. And it also fitted well with the then widely current epistemological slant that scientific theory is fashioned primarily in response to persistent problems reported by experimentalists. By now we know better, not least because of Einstein's rebuttals, found in the Archive, and in books such as my *Thematic Origins* (1973) and John Stachel's *Einstein, B to Z* (2002).

From the use made of the Archive in clarifying old and new speculations about Einstein's way of thinking, one may point to just a very few revealing ones. Einstein's three epochal papers of 1905—sent to the *Annalen der Physik* at intervals of less than eight weeks—seem to be on entirely different fields: the quantum theory of light, Brownian motion, and relativity. It had been suspected by a few that these papers might have arisen from thinking about *one* general problem. An unpublished

---

4. "On the Origins of the Special Theory of Relativity," *American Journal of Physics*, 28, no. 7 (October 1960): 627–636.

letter of January 17, 1952, from Einstein to Max von Laue, confirmed that suspicion. Einstein had already known that Maxwell's theory led to the wrong prediction of the trembling motion of a delicately suspended mirror "in a Planckian radiation cavity." His preoccupation with fluctuation phenomena was at the bottom of all three papers—the consideration of Brownian motion, of the quantum structure of radiation, of his more general reconsideration of what he called "the electromagnetic foundations of physics" itself (as he put it in the "Autobiographical Notes," p. 47). Thanks to letters such as the one to von Laue, one can also retrospectively verify other indications Einstein himself had given of how his mind worked. My favorite example has been his remarkable letter of May 7, 1952, to his old friend Maurice Solovine.[5]

The Continuing Harvest

Just in time to get into the first of the published volumes of Einstein's *Collected Papers*, there came into the Archive a batch of some fifty letters that have become famous as the "Love Letters" between those two young classmates at the Polytechnic Institute in Zurich, Mileva Marić and Einstein. The letters are indeed full of romance and passion, on both sides; but they are even more—a new keyhole for watching young Einstein develop his ideas, over several years, leading to the relativity theory. We had known of his early interest in electromagnetism, a topic he had to study by himself, mainly from the books of Helmholtz, Hertz, Paul Drude, and August Föppl. Jean Pelseneer of Brussels had found Einstein's first essay on that topic, composed at age sixteen or so as a birthday present for his favorite uncle, Caesar Koch. In the newly found letters was more proof of how early he started on the road to relativity.

The letters to and from Albert and Mileva during a six-year span show first of all how these two classmates at the Swiss Polytechnic Institute began to fall in love among the test tubes and lab equipment. Thus in an early letter (March 1899), Albert writes that he realizes "how closely intertwined our psychological and physiological lives are." But

---

5. See chap. 2, "Einstein's Model for Constructing a Scientific Theory," in my book, *The Advancement of Science, and Its Burdens* (Cambridge, MA: Harvard University Press, 1998).

he then adds, "My musings about radiation are now beginning to reach more solid ground—I myself am curious whether something will come of it." As many such letters show, his love life and his physics research are seemingly unified. Thus he writes to Mileva in August 1899 about his admiration for her, but even without starting a new paragraph, his very next, historically important sentences are:

> I have returned the Helmholtz volume [1895] and now study once again Hertz's propagation of the electric force [1892] with great care. . . . I am coming more and more to the conviction that the electrodynamics of moving bodies [note the phrase!], as currently presented [by Hertz], does not correspond to reality, but rather lets itself be presented more simply. The introduction of the word 'Ether' in the electric theories has led to the conception of a medium of whose motion one may talk, without, I believe, connecting with that assertion a physical sense. I believe that electric forces can be directly defined only for empty space."

All this, six years before his relativity paper of 1905, which used as its title the very words in this letter. Here we not only see Einstein dismissing the ether in a curt phrase, but we also get a good idea what books and articles he was reading as he was working toward his breakthrough of 1905.

Among the countless other direct and indirect results of these first years of putting together the Archive, I will just mention one: On October 2, 1962, Thomas S. Kuhn wrote Helen and me a letter asking us to provide examples of how we were cataloging the Archive, to serve as a model for ordering the Niels Bohr Archive in Copenhagen. Einstein and Bohr would have been amused by that turn of events.

It is a pitiful irony that Helen, in part because of the delays caused by the Executor of the Will, Otto Nathan, did not live to see the publication of the first volume of the projected thirty-volume series of Einstein's *Collected Papers*. She died suddenly on February 10, 1982—just six weeks after she had overseen the transfer of the Einstein Archive to Jerusalem. To the end she was in full possession of her lucid faculties. From the late 1960s on, I and others had urged her to fulfill her plan of moving to Jerusalem, where she had relatives and where the Einstein Archive was to find its final home. Perhaps she would have liked to continue from time to time to attend to the papers she had so lovingly dealt with through most of her life, if they were to be transferred earlier. The Hebrew University offered her a room at the Library and living accommodations. She could have added much to illuminate some of the newly acquired materials in the ever-growing Archive. But it was not to be.

Yet, having spent so many years as an "outsider" in the world of science, now she lives on, inside the *Collected Papers*.

— 3 —

Werner Heisenberg and Albert Einstein

To understand better Heisenberg's enormous talent and his responses to the challenges of history, it is useful to examine his deeply significant relationship with another major scientist.

Being Captured by Einstein

At the center of this case are Heisenberg and Albert Einstein. My interest in their interaction was aroused at a December 1965 UNESCO conference on Einstein's work, where I had a first, accidental encounter with Heisenberg himself. I had been invited to lecture on Einstein's epistemology, focusing on his pilgrimage from an early positivism, strongly influenced by Ernst Mach, to what he called a "rationalism."[1] I had followed that change in Einstein's thoughts through reading his correspondence in the files, as described in Chapter 2.

On finishing my lecture, I left the podium, the next speaker came forward, and we met midway. It was Heisenberg. In passing, he whispered to me, "We must talk afterwards." I shall return to this encounter later.

Among the main sources for what follows are Heisenberg's eloquent books and autobiographical articles, the unpublished transcripts of the twelve interviews he gave to the History of Quantum Mechanics Project, his unpublished letters to Einstein, and some thoroughly researched biographies. From these it emerges that in the history of mod-

---

1. That "pilgrimage" is described in chap. 7, "Mach, Einstein, and the Search for Reality," in G. Holton, *Thematic Origins of Scientific Thought: Kepler to Einstein* (Cambridge, MA: Harvard University Press, 1988 ed.), pp. 237–277.

ern physics no one but young Werner was so destined by the fates to be captured by Einstein's relativity theory. In his *Gymnasium* days, he read and loved Einstein's newly published popular book on special and general relativity. He would have been not quite eighteen when he heard of the sensational November 1919 eclipse expedition results. At the University of Munich, where he studied under the guidance of Arnold Sommerfeld, he attended Sommerfeld's lectures on relativity. Heisenberg was also captivated by Hermann Weyl's book, *Raum-Zeit-Materie*. To top it off, one of his closest friends in Munich was Wolfgang Pauli, who, while still a fellow student, was writing his *Handbuch* monograph on relativity theory. When Heisenberg moved to the University at Göttingen, he got more relativity theory from Max Born. In short, it came to him from all sides. Although Pauli wisely warned him to devote his future research to quantum physics instead of relativity, there was no way Heisenberg could escape being fascinated by Einstein's work.

In his years at Munich University, Heisenberg went with some friends on a long bicycle tour around Lake Walchensee. At one point, while they were resting, the talk turned to Sommerfeld's relativity course. Heisenberg was especially struck by a remark from friend Otto Laporte, recalling it later as follows:

> We ought to use only such words and concepts as can be directly related to sense perception. . . . Such concepts can be understood without extensive explanation. It is precisely this return to what is observable that is Einstein's great merit. In his relativity theory, he quite rightly started with the commonplace statement that time is what you read on a clock. If you would keep to such commonplace meaning of words, you will have no difficulties with relativity theory. As soon as a theory allows us to predict correctly the result of observations, it gives us all the understanding we need.[2]

This "instrumentalist" or "operational" view of Einstein's method was quite common at that time, and for decades afterwards. As we shall see below, Laporte's long-remembered praise of it laid the groundwork

---

2. Heisenberg, *Der Teil und das Ganze* (Munich: R. Piper and Co., 1969), p. 49. My translation.

for one of Heisenberg's key insights many years later, which changed physics forever.

In the summer of 1922, Sommerfeld arranged for Heisenberg to go to Leipzig, where Einstein was to give a lecture. It was to be Heisenberg's first encounter with Einstein, but instead it turned into a surrealistic glimpse of things to come. As Heisenberg entered the crowded lecture hall, a handbill was forced on him, signed by the Nobel physicist Philipp Lenard and eighteen other German scientists. It contained a vicious attack on Einstein, whose theory, as Heisenberg recalled, "was said to be nothing but wild speculations, alien to the German spirit, and blown up by the Jewish press."[3]

Heisenberg was shaken by this political attack on scientific truth—so much that he didn't even notice that the speaker on the distant platform was not Einstein but rather Einstein's courageous friend and colleague Max von Laue. Einstein had decided not to come, knowing that he was in mortal danger from Nazi rowdies.

"Only the Theory Decides"

The first real meeting between our two protagonists occurred in 1924, when Einstein—at age forty-five about twice as old as Heisenberg—came briefly to Göttingen. The recent work of Bohr, Kramers, and Slater—the BKS theory—was hot news. But because it relaxed the requirements of strict causality and of energy and momentum conservation, Einstein wrote to Max Born that if this kind of science would persevere, "I would rather be a shoemaker or employee in a gambling casino than a physicist."

Against that background, Einstein and Heisenberg had a private talk in 1924, during a walk through the neighborhood. (By the way, what has happened to the life of scientists? Where have all those walks gone?) As Heisenberg, a proponent of Bohr's point of view, immediately wrote to his parents, "Einstein had a hundred objections." Coming from a scientist whose work Heisenberg had been admiring since early youth, this rejection of the new way of doing physics must have been difficult. But he consoled himself, as he said in one of his later interviews, that his

3. Ibid., p. 67.

generation, having "grown up into a complete mess" in quantum physics, was in the happy position of being able to give up old schemes if necessary.

On September 25, 1925, Heisenberg published in the *Zeitschrift für Physik* his brilliant breakthrough to quantum mechanics, "On the Quantum Theoretical Reinterpretation of Kinematic and Mechanical Relations." He had arrived at it during two lonely weeks on the island of Helgoland, to which he had fled to recover from hay fever. The abstract of the paper announced Heisenberg's fundamental guiding principle: to restrict oneself to observable properties of a spectrum, eschewing models built on unobservables such as the position and periods of electrons in the atom. As he put it, "This work is an attempt to find foundations for a quantum-theoretical mechanics which is based exclusively on relations between quantities that are in principle measurable."

Heisenberg later observed that his crucial insight was an echo from the days when he had been struggling with relativity theory at the University in Munich. In his work leading up to that 1925 paper, he remembered the philosophy presented as Einstein's viewpoint by his friend Otto during that bicycle tour: "regard only the observable magnitudes as the indication of atomic phenomena."[4]

If Heisenberg had any illusion that his article would be approved by Einstein, he was wrong. One of Heisenberg's five surviving letters in the Einstein Archive, dated November 30, 1925, is evidently a reply to a note from Einstein (now lost) that had contained many objections. In his response, Heisenberg tried to hold out the hope of an eventual peaceful bridging between Einstein's theory of light quanta and what he called "our quantum mechanics." Heisenberg also drew prominent attention to his having used only "observable magnitudes" in his theory. All to no avail.

The following year, 1926, is one of high drama in this growing but troubled relationship. In April, Heisenberg gave a two-hour lecture on his matrix mechanics in von Laue's famous physics colloquium at the University of Berlin. In the audience, with a whole group of potentates, was Einstein. It was their second meeting. Einstein, interested and no doubt disturbed by the lecture, asked Heisenberg to walk home with

---

4. Ibid., p. 88.

him (there is that walk again) and thus ensued a remarkable discussion, which Heisenberg first reported in print in 1969.

In that discussion with Einstein, Heisenberg once more tried to draw attention to his having dealt not with unobservable electron orbits inside atoms, but rather with observable radiation. He reports having said to Einstein: "Since it is acceptable to allow into a theory only directly observable magnitudes, I thought it more natural to restrict myself to these, bringing them in, as it were, as representatives of electron orbits." Einstein responded, "But you don't seriously believe that only observable magnitudes must go into a physical theory?" Heisenberg goes on, "In astonishment, I said, 'I thought that it was exactly you who made this thought the foundation of your relativity theory. . . .' Einstein replied, 'Perhaps I used this sort of philosophy; but it is nevertheless nonsense [*Unsinn*].'" And then came Einstein's famous sentence: "Only the theory decides what one can observe."[5]

All this must have come to Heisenberg as a scathing attack on what he regarded as his fundamental orientation, derived from reading Einstein's early works and guided by them from the start, right through to his most recent triumph. Einstein, whose development away from positivistic instrumentalism had escaped Heisenberg's notice, went on to explain at length how complicated any observation is in general, how it involves assumptions about phenomena that in turn are used in theories. For example, one almost unconsciously uses Maxwell's theory when interpreting experimental readings involving a beam of light.

Perhaps this discussion helped Heisenberg eventually to embark on his own epistemological pilgrimage, which ultimately ended with a kind of neo-Platonism in the description of nature through the contemplation of symmetries. But in 1927, just before starting on his next breakthrough—later called the uncertainty principle paper—Heisenberg suddenly remembered Einstein's provocative statement, "Only the theory decides what one can observe." It was a key to Heisenberg's ad-

---

5. Ibid., pp. 91–92. In his 1829 *Course de philosophie positive,* Auguste Comte had already noted that description of observations cannot be free from the prior acceptance, tacit or not, of a conceptual scheme. He wrote that while "every theory has to be based on observation, it is, on the other hand, also true that our mind needs a theory in order to make observations."

vance. As he put it later in one of his interviews, "I just tried to turn around the question according to the example of Einstein."

At this point I should pause briefly to return to the unfinished story of my own encounter with Heisenberg in 1965. After giving his lecture, Heisenberg came over to tell me in detail about that 1926 meeting with Einstein, and what it had meant for him—all this four years before he published anything about it. Indeed, as if to make sure I had it straight, Heisenberg followed up by sending me a letter in January 1966, in which he repeated the account and added a rather striking conclusion: While the theory determines what can be observed, the uncertainty principle showed him that a theory also determines what *cannot* be observed. Ironically, Einstein, through the 1926 conversation, had provided Heisenberg with some genetic material for the creation of the uncertainty principle article of 1927.

Descending along Two Tracks

We can now follow the effect of Einstein on Heisenberg along two diverging tracks. Both start at a high level but descend eventually into terrifying terrain below. One track is the scientific one. Despite all his misgivings, Einstein of course realized the brilliance of Heisenberg's work. He nominated Heisenberg for a Nobel Prize for three years before Heisenberg was so recognized, even though Einstein to the end believed that Heisenberg's way of doing physics would ultimately turn out not to be true to the thoughts of the "Old One," the Creator.

The third meeting of the two men took place in October 1927, at the six-day-long Solvay Congress of physicists in Brussels. That conference was the scene of famous debates, mainly between Einstein and Schrödinger on one side and Bohr, Heisenberg, and their like-minded colleagues on the other.[6] It soon became clear that the Copenhagen spirit had triumphed. Day after day, Einstein presented ingenious arguments, which Bohr then answered before nightfall, until Paul Ehrenfest finally said, according to Heisenberg, "Einstein, I am ashamed for you."

6. The canonical reference to these debates is Niels Bohr's account, "Discussions with Einstein on Epistemological Problems in Atomic Physics," chap. 7 in Paul A. Schilpp, ed., *Albert Einstein: Philosopher-Scientist* (Evanston, IL: The Library of Living Philosophers, Inc., 1949).

Heisenberg, in a later interview, added a shrewd point: "I would say that a change had taken place, which I can only express in terms of lawsuits. That is, the burden of proof was reversed.... That made a complete change of view among the younger generation." Ironically, the same kind of reversal of fortunes had happened long before, in the triumph of Einstein's relativity over his opponents. But Heisenberg's last surviving letter to Einstein, written a few months before the Brussels meeting, already showed the cocky self-confidence of the victors in that new struggle. Heisenberg wrote that while in the new quantum mechanics Einstein's beloved causality principle was baseless, "We can console ourselves that the dear Lord God would know the position of the particles, and thus He could let the causality principle continue to have validity."

Strangely enough, in 1954, a year before Einstein died, Heisenberg sought out Einstein once more. Meeting with him in Princeton, Heisenberg found that Einstein's view had not changed since the 1927 Solvay Congress. Despite all Heisenberg's persuasive skills, Einstein just said, "No, that's nothing. That's not the thing I am after. I don't like your kind of physics. I think you are all right with the experiments . . . but I don't like it."

The second track, which follows the later relation between the two men, concerns the full emergence, as from the mouth of the beast, of a soul-destroying hatred that had been building in Germany since the early 1920s. For a time, Heisenberg continued to mention Einstein in his lectures and publications, but by the 1930s the scene was dominated by demons. Raving articles were published in 1935 by Johannes Stark that branded Heisenberg the "spirit of Einstein's spirit." The attacks on Heisenberg and on theoretical physics as such culminated on July 15, 1937, with an article in the official journal of the SS, *Das Schwarze Korps*. That article, endorsed by Stark, called Heisenberg a "white Jew" and dismissed relativity and quantum theory as non-German, Jewish thinking.

There followed a year-long attempt by Heisenberg to obtain exoneration from Heinrich Himmler, head of the SS, who happened to be a family acquaintance. That effort finally succeeded, although with the stipulation that in the future Heisenberg would "clearly separate for your audiences, in the acknowledgment of scientific research results, the personal and political characteristics of the researcher." Privately, Himmler had his eye on Heisenberg as a possible researcher on Himmler's own crazy obsession, the "World Ice Theory," of which more in Chapter 11. But any future playwright constructing a version of the Heisenberg-Einstein relation will not be able to avoid including the cries, off stage and ever more distant, of the unmentioned millions who had also loved their homeland but had no way to make a deal with Himmler, or to bribe an SS man bent on murder.

Recasting the Portrait of Einstein

Years later, at last in peacetime, Heisenberg was securely installed as the leader of a new generation of German physicists—as he had hoped to be all along. But now, in two of Heisenberg's lectures, we find passages that signal the depth to which his relationship with Einstein had fallen.

Shortly after Einstein had died in 1955, Heisenberg published a popular article entitled "The Scientific Work of Einstein."[7] The essay began with a generous assessment of Einstein's contributions but then found a serious fault with him, namely "that Einstein, to whom war was hateful, should have been moved by the infamous practices under Nazism to write a letter to President Roosevelt in 1939, urging that the United States vigorously set about the making of atomic bombs" which eventually "killed many thousands of women and children."[8]

That bitter statement was at the very least a major exaggeration. The famous letter of August 1939 that Einstein signed had been written just as the German war machine was poised to start its Blitzkrieg—and, as we now know, four months *after* Paul Harteck and Wilhelm Groth had asked the German War Office to investigate nuclear explosives. Far from urging that the United States vigorously set about the making of atomic bombs, Einstein's letter was, in his own words, "a call for watchfulness and, if necessary, quick action," not least because "Germany has actually stopped the sale of uranium from the Czechoslovakian mines which she has taken over."

---

7. In *Universitas*, 10, no. 9 (1955): 8978–902; reprinted in his book *Across the Frontiers* (New York: Harper and Row, 1974), a book first published as *Schritte über Grenzen* (Munich: R. Piper and Co., 1971).

8. *Across the Frontiers*, p. 6.

The letter asked only to establish a liaison between the U.S. government and the physicists and for help in raising funds for experimental work in university laboratories, if necessary from private donors and industrial laboratories. The direct result was that all of $6,000 was made available to Fermi and Szilard at Columbia University. Einstein declined the invitation to be a member of a group to coordinate further research.

Einstein signed a second letter to Roosevelt in March 1940, reporting that he had heard that research on the use of uranium was indeed going on in Germany; this letter, too, produced little action. In fact, the U.S. government did not gear up seriously until October 1941, when it received the so-called Maud Committee report, with the conclusions of a British-sponsored study on how one might produce an atomic bomb.

Leo Szilard persuaded Einstein to write a third letter, in 1945, that was simply a letter of introduction to Roosevelt for Szilard, who was not allowed to tell Einstein the need for it. Szilard hoped to use this letter to convey to Roosevelt his doubts on "the wisdom of testing and using bombs." But Roosevelt died before this plea reached him.

Einstein himself, now regarded as "unreliable" by the authorities, was carefully shielded from direct knowledge of the Allied nuclear project. This secrecy even resulted in a moment of comedy. In late December 1941, Vannevar Bush tried to get advice from Einstein on building uranium hexafluoride diffusion plants. Because Einstein was given only vague details, however, his reply was useless. Bush then was asked by the intermediary if Einstein could be given more information. Bush cried, no, don't tell him one more thing, or he will guess the rest of the project, and then he might blab. The voluminous files the FBI kept on Einstein show that director J. Edgar Hoover was personally devoted to having Einstein spied upon. It is ironic that on one side of the Atlantic, Heisenberg was called a "white Jew" while Einstein, on the other side, was considered by some to be a red one.

Heisenberg's (1955) remarks about Einstein were not to be an isolated exaggeration. Heisenberg gave a second, more detailed attack on Einstein in June 1974, when he spoke, of all places, in the so-called Einstein house in Ulm, Germany. As in 1955, he began with a generous survey of Einstein's work on relativity. He then repeated some of the points made in earlier publications, including an account of Einstein's rejections of Heisenberg's theories. At that point, Heisenberg had to add something, "in order not to leave the portrait of Einstein all too incomplete." Einstein, he said, "wrote three letters to President Roosevelt, and thereby contributed decisively to setting in motion the atom bomb project in the United States. And he also collaborated actively, on occasion, in the work on this project."[9]

If there is ever to be a play based on the relation between these two men, the playwright will perhaps note that these astonishing exaggerations, uttered in Einstein's birth town, were delivered in a lecture with the title "Encounters and Conversations with Albert Einstein." In that last talk, Heisenberg, two years before his death, had his final encounter with the person whom he had once called his *Vorbild,* his model; the person who for good and ill had unknowingly been the cause both of deep insights and of fierce insults throughout Heisenberg's scientific and personal life; and whose acceptance Heisenberg had sought again and again, always in vain. Those rejections by Einstein, at each encounter, had evidently left on Heisenberg a lasting scar, which bled once again at that final lecture. Niels Bohr, to his death in 1962, was also deeply saddened by Einstein's constant refusal to accept his interpretation and program. And as for Einstein, he often cursed the quantum he himself had set loose, only to have it haunt him in the form of a physics, initiated largely by Bohr and Heisenberg, that he could not accept.

In that future play, as the curtain falls on these three extraordinary men, even the evil spirit that has been watching them from the wings of the stage, and that had haunted the whole terrible century, will, in the end, have to shed a tear for humanity.

---

9. W. Heisenberg, *Encounters with Einstein* (Princeton: Princeton University Press, 1989), p. 120.

4

Bohr, Heisenberg, and What Michael Frayn's *Copenhagen* Tries to Tell Us

Werner Heisenberg's unexpected visit to Niels Bohr, in September 1941 in German-occupied Denmark, was a non-event in terms of science or of history. During the 1920s, the course of physics had been changed as a result of the close interactions between these two great scientists. This time it was not. Nor had this encounter in Copenhagen, at the height of the German Army's successes throughout Europe, any known effect on the course of the war or on the ongoing work of the nuclear scientists on either side at the time. As far as we know, whatever Heisenberg may have hoped to gain from this meeting did not occur.

Copenhagen

But on this unfruitful base, imaginations from various sides have built castles full of mystery and drama. One of these is Michael Frayn's play *Copenhagen*. The author has cleverly highlighted the concept of uncertainty, associating it with capacious speculations of what may have actually happened in that meeting. The choreography of the actors' movements, apparently meant to evoke the supposed motions of electrons in Bohr's first atomic theory, adds to the theatrical experience. The dramatist's device to start the action over and over again from the same point, as had been done earlier so successfully in Max Frisch's *Biographie*, adds a hypnotizing element.

There are only three actors on the stage, representing versions of Bohr, his wife Margrethe, and Heisenberg. But they have become ghosts, recounting the extraordinary scientific breakthroughs in quantum mechanics, with asides about other remarkable physicists with whom they had collaborated or fought. Often they become suddenly unaware of one another's presence, adding to the spooky quality of their interaction. But again and again these three return to two haunting puzzles. One is why the Germans never succeeded in building an atomic bomb. The other, perhaps connected with the first puzzle, is why Heisenberg had suddenly turned up at Bohr's doorstep:

> *Margrethe:* Why did he come? What was he trying to tell you?
> *Bohr:* He did explain later.
> *Margrethe:* He explained over and over again. Each time he explained it became more obscure . . . I've never seen you as angry with anyone as you were with Heisenberg that night.
> *Heisenberg:* Now we're all dead and gone, yes, and there are only two things the world remembers about me. One is the uncertainty principle, and the other is my mysterious visit to Niels Bohr in Copenhagen in 1941. Everyone understands uncertainty. Or thinks he does. No one understands my trip to Copenhagen.

I was not surprised that Frayn won Broadway's coveted Tony Award for the best play of the year 2000. And like many scientists, I hope that a piece for the theater may create interest among some of the public in Bohr's and Heisenberg's science—although for most viewers, the discussion on the stage is often so elusive and mysterious that, in the words of a reviewer,[1] "'Copenhagen' makes you feel smarter for having seen it . . . even if you don't really understand it."

But by mixing in one work the three quite different worlds of science, history, and theater, it is highly likely that much of the audience will confuse the play—a work of fiction—with a historical documentary. So many think they know all about Mary Stuart or J. Robert Oppenheimer from having read or seen the dramas using those names in the titles. They tend to forget that the task of the poet and dramatist is, as Samuel Taylor Coleridge put it in his *Biographia Literaria* of 1817, to create in the reader or viewer the "willing suspension of disbelief" and "poetic faith." John Keats, at about the same time, also memorably celebrated what he called the "Negative Capabilities" of great authors, which he defined as their ability to remain "content with half knowledge."[2]

---

1. *New York Times,* June 2, 2000, p. B6.
2. Grant F. Scott, ed., *Selected Letters of John Keats* (Cambridge: Harvard University Press, 2002), p. 60.

Jungk, Powers, and the Start of a Legend

These truths are especially relevant for the case at hand; for Frayn based his play—as he was writing it originally for publication in Great Britain in 1998—on two deeply flawed publications by journalists. The first is Robert Jungk's book, *Brighter Than a Thousand Suns: A Personal History of the Atomic Scientists* (New York: Harcourt Brace, 1958), based on his German edition of 1956. In it, Jungk published the most familiar version Heisenberg himself gave of what in Frayn's play becomes the key event—the private conversation between Heisenberg and Bohr during that evening in 1941, which Bohr, disturbed by something Heisenberg had said, ended abruptly.

Heisenberg's own account came to Jungk in a four-page letter of January 18, 1957, in response to Jungk's request of December 29, 1956. Heisenberg's account, which Jungk published in his book only in part, asserted that in 1941 the researchers in Heisenberg's *Uranverein*—the organization of German scientists assembled soon after the outbreak of World War II to put nuclear physics to use in the war effort—"knew that fundamentally one could produce atom bombs, but we estimated the necessary technical expenditure as larger than it was then in fact" (my translation, from Heisenberg's German original). Still, Heisenberg contended, physicists engaged in such work could have "decisive influence on further developments, since they could argue that by means of extreme effort it would perhaps be possible after all to put them into play." And his discussion with Bohr, Heisenberg noted in his letter to Jungk, "probably started with a question I raised in passing whether it was right for physicists to devote themselves in wartime to the uranium problem." Heisenberg observed that Bohr "understood at once the implication of this question" and was shocked *(erschrocken)* by this train of thought, assuming "that I had intended to convey to him that Germany had made great progress in the direction of manufacturing atomic weapons." Heisenberg added he was unable to "correct this false reaction."[3]

3. Quoted (in my translation) from the revised edition of Robert Jungk, *Heller als tausend Sonnen: Das Schicksal der Atomforscher* (Munich: Wilhelm Heyns Verlag, 1990), pp. 399–401. Heisenberg's letter was rather freely translated in Jungk's English-language edition, mentioned above.

There are significant parts of Heisenberg's letter which Jungk chose not to print and which, as far as I know, strangely seem not to have been made public. For example, in his immediate next sentence Heisenberg reported he told Bohr of the "moral consideration" of making atomic weapons. Heisenberg had raised the same theme in a previous letter to Jungk (November 17, 1956). In it Heisenberg confessed to have suppressed any memory of an earlier discussion about atomic weapons; "perhaps out of fear it was pushed away inside [*vielleicht aus Angst innerlich verdrängt*]," and Heisenberg elaborated on his curious experience of *Verdrängung*.

To be sure, even in the parts of his letter which Jungk did publish, Heisenberg had been careful to warn at the beginning about the uncertainty of his memory: "In my memory, which naturally after such a long time may deceive me . . ." He also introduced the most controversial part of his recollection of the encounter with Bohr with the words "The conversation may have begun . . ." Nevertheless, Heisenberg's account, together with Jungk's report in his widely read book on his own conversations and correspondence with other German nuclear scientists, provided the key suggestion which Frayn's play vastly expanded: that at least for Heisenberg, *an impeding moral compunction may have existed about his working toward an atomic bomb.*

This was not the first time this idea had been launched. It had surfaced on August 7, 1945, in a remark by C. F. von Weizsäcker to Heisenberg and the other German scientists who were then being detained in England at Farm Hall, with their conversations secretly recorded by their captors. Immediately after those scientists had heard about the existence and use of the atomic bomb on the part of the Allies, von Weizsäcker proposed: "History will record that . . . the peaceful development of the uranium engine [reactor] was made by the Germans under the Hitler regime, whereas the Americans and the English developed this ghastly weapon of war."[4] (To von Weizsäcker's credit, in 1988, in his revealing autobiographical memoir, he confessed at last that the

---

4. For the reports on the conversations and detailed commentaries, see Jeremy Bernstein, *Hitler's Uranium Club: The Secret Recordings at Farm Hall*. The scientist's quoted passage appears on p. 138 of the book's paperback edition (New York: Copernicus Books, 2001).

Uranverein scientists had been hoping to build a bomb.[5] And to Otto Hahn's credit, at Farm Hall he responded to von Weizsäcker at once: "I don't believe that"—which nobody else there contradicted.)

Max von Laue was present at the time, having strangely been made part of the group of internees, although this distinguished physicist had been an outspoken opponent of the Nazi regime throughout and had not been engaged in atomic energy research. In a later letter, von Laue recorded his observation that right then and there, at those August 1945 conversations *(Tisch-Gespräche)*, a *Lesart* (a version) of history was fashioned: the German failure to produce a bomb was not chiefly due to the numerous errors[6] made during their research, together with insufficient help obtained from their administrators and the ultimate disillusion of the latter with the whole project. Also, it was not, as now most agree, that compared with the Allies the German researchers had a relative "lack of zeal" about their work. For one of the major mistakes the German scientists had made was to underestimate the likelihood of the other side to have the skill and wits to pursue the bomb project successfully. For example, hearing on August 6, 1945, on the radio about the first atomic bomb, they felt at first it was a hoax, a *chemical* explosion, with Heisenberg saying, according to the Farm Hall Papers, "All I can

---

5. Carl Friedrich von Weizsäcker, *Bewusstseinswandel* (Carl Hanser Verlag, 1988), pp. 362–383. On p. 365, a *Stern* interviewer asks the author: "*Die Idee war also: Wir wollen die Bombe bauen, damit wir etwas in der Hand haben? C. F. v. W.: Ja. Oder mindestens, wir wollen so nahe an die Bombe herankommen, wie wir eben können.*" ("So the idea was: We want to build the bomb, to have something in hand?" C. V. v. W.: "Yes. Or at least, we want to come as close to the bomb as we could.")

6. See for example Bernstein's book. (Max von Laue's letter is reprinted there, in Appendix B.) The book is an excellent source for understanding the ambitions of Heisenberg's Uranverein and the many reasons for its ultimate lack of success during more than five years of work to design and build even one functioning reactor. Also see David Cassidy's article, "A Historical Perspective on Copenhagen," *Physics Today 53*, 7 (2000): 28–32, accessible via the Internet under http://www.aip.org/pt/vol-53/iss-7/p28.html, and Rudolf E. Peierls, *Atomic Histories* (Woodbury, NY: American Institute of Physics, and New York: Springer Verlag-New York, 1996), pp. 108–116. Peierls analyzes some of the decisive errors made during the German nuclear program. Heisenberg himself, in an interview with J. J. Emerec (August 29, 1967), somewhat harshly criticized a crucial "error" Walther Bothe had made in "a measurement of the neutron absorption coefficient of pure carbon"—a crucial measurement, but one that Heisenberg said had not been checked by repetition.

suggest is that some dilettante in America who knows very little about it has bluffed them."[7] Rather, that failure was now to be traced conveniently either to the unrealistic timetable for achieving it or, as von Laue heard it from von Weizsäcker, "because they simply did not want to have it at all [*weil sie überhaupt nicht wollten*]." This last phrase again underlined the suggestion of a moral reluctance to do such work, in contrast with the eagerness of the Allies.

That story of a supposed moral contrast between the opposing sides was later elaborated by some German scientists, and it undoubtedly helped them at home to explain to the public their failure during nearly six years of work on the project. In addition, the same story also was given a prominent place in Robert Jungk's book because it fitted perfectly Jungk's own agenda at the time, which he frankly revealed in his first chapter, entitled "How the Book Came to Be," as published in the German edition. Indeed, Jungk's correspondence shows that he was at first almost grotesquely grateful for his "coup." But eventually he came to understand that he had been gravely misled—even, as he put it, "*verraten*," betrayed, having been used to propagate *"eine Legende."*[8]

The legend might eventually have blown away for lack of credible evidence. But it was given new, vigorous life by the widely distributed publications of a second journalist, Thomas Powers. In his book *Heisenberg's War* and his many articles on the same topic, Powers took the tale of moral compunction to its logical extreme: Now Heisenberg's failure as a leader of the Uranverein during the war years was portrayed as essentially an act of *conscious sabotage*.[9] Heisenberg was portrayed as fully understanding what had to be done but keeping it secret, misleading all his co-workers throughout, "giving a different estimate for criti-

---

7. Bernstein, *Uranium Club*, p. 116, n. 3. On several of these points, there are also two other relevant sources: Jonothan Logan, "A Strange New Quantum Ethics," *American Scientist*, 88 (July–August 2000): 356–359, and Abraham Pais, "What Happened at Copenhagen?" *Hudson Review* (Summer 2000): 182–189.

8. For a detailed presentation, see Robert Jungk's autobiography, *Trotzdem: Mein Leben für die Zukunft* (Munich: Carl Hanser Verlag, 1993), pp. 297–300.

9. Thomas Powers says he never used the word *sabotage*. He prefers the circumlocution that Heisenberg "found a way of leading it [the bomb project] into a closet where it languished for the remainder of the war" (*The New York Review of Books*, April 11, 2002, p. 85).

cal mass to different people"[10] and effectively subverting the Germans' path to developing an atomic weapon. As Powers put it: "Heisenberg did not simply withhold himself, stand aside, let the project die. He killed it."[11] To this day, this and similar versions are being energetically defended by a few, even though Powers has acknowledged that he has not convinced any historians.

But that brings us back to the play *Copenhagen*, because Frayn revealed in the original "Postscript" (Postscript I, 1998), printed in the first edition of his play's published text, that it was Powers's "extraordinary and encyclopaedic" book which "first aroused my interest in the trip to Copenhagen."[12]

The play skillfully prepares the audience for the moving climax near the end: Heisenberg accuses Bohr of having gone to Los Alamos, upon which Bohr is made to add: "To play my small but helpful part in the deaths of a hundred thousand people," whereas Heisenberg "never contributed to the death of one single solitary person." At this, Frayn's Heisenberg exults: "There'd be a place in heaven for me," implying that Bohr belongs in the other place. That sentiment is immediately reinforced by Heisenberg's final bravura soliloquy in which he bemoans what the ferocious Allies (without provocation?) had done to his beloved homeland during the war.

As the audience was leaving the performance in New York at its U.S. premier in Spring 2000, I saw tears in many eyes. At least for those persons, all "uncertainties" had given way to a "knowledge" of what really happened on that day in September 1941 and to which side the moral victory belonged. The triumph of good fiction was palpable; its success

---

10. Ibid.

11. Thomas Powers, *Heisenberg's War: The Secret History of the German Bomb* (New York: Knopf, 1993), p. 479. For typical, authoritative presentations of the assumptions and errors in Powers's book, see the book review in *American Historical Review*, (December 1994): 1715–1718, and R. E. Peierls, *Atomic Histories* (Woodbury, N.Y.: A.I.P., 1997): 108–116.

12. Michael Frayn, *Copenhagen* (London: Methuen Drama, 1998), p. 100; a second edition with a few changes but a completely revised Postscript was published in 2000 (New York: Anchor Books). That revised Postscript (Postscript II) is available on the Internet (http://web.gsuc.cuny.edu/ashp/nml/copenhagen). A "post-postscript" (Postscript III) was circulated at a meeting in Washington, April 2002.

even made plausible the implication that it correctly presented the actual historical events as well.

Bohr's Refutation of the Legend

As to historical facts, the account of the famous meeting in September 1941, as offered by the actual Heisenberg and Jungk, presents only a one-sided version. At the time of the New York performance, Bohr's own reaction in writing was still kept secret. Only since February 2002 has it been freely accessible to all (see www.nba.nbi.dk). But as it happened, in 1985, at a meeting in Copenhagen in honor of Niels Bohr's memory, I was approached by Bohr's son, Erik Bohr, probably because of my involvement in organizing several archives of scientists. He showed me an unsent letter which he explained had been written by his father but found after Bohr's death, folded into Bohr's copy of the book by Jungk. That striking letter (the first, longest, and most important of the documents released later), addressed to Heisenberg, is, as intended, a powerful corrective to previous stories disseminated by Heisenberg, Jungk, and others. It takes serious issue with every detail of Heisenberg's published version of the 1941 meeting, in quite firm language—so firm that this may have been one of the reasons Niels Bohr apparently decided not to mail it.

When asked what should be done with the document, I advised that it be preserved and put into the Niels Bohr archives. Today that letter is there, together with the ten other drafts of it. Although the letter had not yet been embargoed by Bohr's family when it was shown to me, I thought it would have been inappropriate for me to say more about it until it was released. So, unless Bohr's family had decided otherwise, the world would have remained for many years with half-knowledge about what happened during that meeting in 1941.

Now that the drafts are available freely on the Internet, every interested reader should take advantage of a rewarding exercise: putting the portion of Heisenberg's letter as published in Robert Jungk's book next to a printout of those fascinating drafts by Niels Bohr. As Bohr intended them, his documents illuminate that meeting with Heisenberg in 1941 and show how incomplete and even erroneous previous accounts were. They also allow us again to see Bohr's mind at work: He typically goes again and again over the same ground in successive drafts, bringing in new details—as he did when dictating his physics papers.

Bohr's documents also remind us that Heisenberg had come to Denmark with his colleague von Weizsäcker, that they had also spoken with others there, and that those encounters in 1941 were also not successful. Thus one of Bohr's drafts records: "During conversations with Møller [the Danish physicist], Heisenberg and Weizsäcker sought to explain that the attitude of the Danish people toward Germany, and that of the Danish physicists in particular, was unreasonable and indefensible since a German victory was already guaranteed and that any resistance against cooperation could only bring disaster to Denmark."

Bohr's first and most interesting document starts by offering Heisenberg the opportunity for excusing himself, by suggesting that Heisenberg's memory might have greatly deceived him when he wrote to Jungk. And in fact, as noted above, Heisenberg had started that letter with the disclaimer: "As far as I remember, although I may be wrong after such a long time . . ." Later Heisenberg had used the word *probably* when trying to describe how his talk with Bohr started; and later still: "I may have replied . . ." In stark contrast, Bohr writes at the beginning: "Personally, I remember every word of our conversations."

Bohr's first document then recounts that Heisenberg in September 1941 freely offered Bohr the astounding confession that he was "completely familiar with them [atomic weapons] and had spent the past two years working more or less exclusively in such preparation." Bohr explains his reaction: not anger, as some insist who want to portray him as an angry old man, but the shock of *fear*. After all, the prospect Heisenberg offered him was that of a successful and energetic pursuit by the German team to make an atomic bomb, at the very time when Hitler's armies were making their greatest advances. That ghastly portent was, as Bohr wrote there, "a great matter for mankind."

Even at that point, Bohr—to whom Heisenberg had been, during their long and fruitful scientific collaboration in the 1920s and early 1930s, a kind of successful son—again gently suggests to Heisenberg a way out. Bohr writes that Heisenberg's report to Jungk that Bohr had been shocked by the idea that atomic weapons were possible in principle was a "misunderstanding . . . due to the great tension in your mind." What Bohr called his "shock" was, he writes, that "as I had to understand it, . . . Germany was participating vigorously in a race to be the first with atomic weapons." To make sure Heisenberg understood properly, Bohr then repeats that his own memory of the conversation was clear. And that is entirely plausible: the crucial discussion between Heisenberg and himself had been very brief; the topic raised by Heisenberg was immensely important; and afterwards Bohr reported on the conversation, in "thorough discussion" with others, including members of his institute and "other trusted friends in Denmark." Not much later, Bohr, having been spirited out of occupied Denmark, spoke with members of the British intelligence who de-briefed him.

In some of the last drafts, Bohr repeats that he "carefully fixed in [his] mind" every word that was uttered when he and Heisenberg met on that ominous occasion. Therefore he finds it "incomprehensible" that Heisenberg should have claimed later he had "hinted" that the German scientists "do all they could to prevent such an application of atomic science." That spin of supposed moral qualms has of course been at the center of revisionist writings.

There is much to ponder here. But, in short, when comparing Heisenberg's letter and Bohr's documents, we see that Bohr contradicts and tries to correct every major point in Heisenberg's published account. In fact, one may speculate that Bohr in the end did not send off his letter to Heisenberg, on which he had worked for so long, because even the (to us) relatively mild words in his documents seemed to him to be uncharacteristically strong. And it is yet another irony that Bohr, who had no reason to hide or misremember anything, eventually did not mail his letter, whereas Heisenberg did let Jungk publish his account, even though when he did so he had good reasons for *Verdrängung*, and for misremembering.

The Playwright as Moral Arbiter

A natural question raises itself. What can be done about the text of the play, written in 1998, based in part on Powers's untenable main contention that Heisenberg knew how to make a workable bomb but kept the knowledge to himself? Frayn confessed in his second postscript of 2000 that only after he had written his play did he read David Cassidy's "excellent biography" of Heisenberg and at last understand also the German scientists' Farm Hall conversations, thanks to Bernstein's commentary. Thus, he now dismissed Powers's main thesis with the cutting comment, "If he [Heisenberg] had kept the fatal knowledge . . . from anyone, as Powers argues, then it was from himself." Worse yet, now that we know from Bohr's documents and von Weizsäcker's own acknowledgment of Germany's project to try to build an atomic bomb, what should the author of the play now do about such passages in which his Heisenberg remarks dramatically: "I understood very clearly. I simply didn't tell the others"? And later, "*I was not trying to build a bomb*"? Perhaps the actor will be instructed to deliver such lines with heavy irony.

Frayn did make some minor changes and corrections for the play's revised version of 2000, mostly cleaning up small errors in the physics and acknowledging other scientists' contributions. But even though it is now acknowledged to be remote from historical reality, the body of the play must of course stand. After all, it remains a hugely successful work of *fiction* for the theater, honored with awards, no matter that a very different story is known about the actual meeting in 1941.

Moreover, the ever-pregnant Muse of History may well have surprises in store for us, in days and years to come. New documents are bound to appear, perhaps details about the slave laborers who had to process the uranium for Heisenberg's Uranium Club. Such findings may keep historians busy but surely should not require Michael Frayn to issue yet more postscripts—as long as he sees his role to be a writer of fiction and not also of a factual documentary, even of one that has a moral message.

Yet, there are signs that he has chosen to step out of the role of a playwright and reveal himself as a moral arbiter between the actual persons involved, rather than only between actors on a stage. He spoke more recently of the audience drawing "its own moral conclusions." And in an interview published in the *New York Times* (February 9, 2002), he goes further: "Heisenberg didn't in fact kill anyone" whereas Bohr "did actually contribute to the death of many people"—referring, respectively, to one person who in fact had been working for many years, with varying degrees of enthusiasm, for Germany's war machine, and to the other who had to flee for his life from Denmark and came late to Los Alamos, where his principal activity was to develop postwar arms control policies.

In this double role the playwright seems sadly to forget that his thing's a *play*. That should be enough. Other plays, imagined as more or less vague semblances to historic events—from Shakespeare's *Richard III* to Brecht's *Galileo*—have survived well and retain their authenticity, despite their grave dissonances with respect to historians' analyses of the actual cases. Let me quote John Keats again. He advised that authors of fiction should be "capable of being in uncertainties, mysteries, doubts, without any irritable reaching after facts and reason."

— 5 —

Enrico Fermi and the Miracle of the Two Tables

There is something quite special about the place of Fermi in history. We all know that in the turbulent flow of time there have arisen, on rare occasions, events that did not fit any previously made plan but nevertheless powerfully shaped all subsequent history. Among the most spectacular examples is of course the discovery by a captain, born in Genoa, who set sail toward Asia but serendipitously encountered instead the land that came to be called the New World. From that moment, the clock for the modern period was set. Another instance of a similar sort was when a then still obscure professor of mathematics at the University of Padua, having used his homemade spyglass for terrestrial explorations, raised it to scan the heavens and was the first to see there the evidence, in the appearance of the Moon, Jupiter, and Venus, that the existing worldview had to be replaced by a new one. That is when the clock for modern science suddenly came alive. And a third example was a seemingly unplanned event that took place in Rome in October 1934, with transforming consequence—for large sections of physics, chemistry, engineering, medical research, and ultimately for politics and warfare.

I refer of course to the discovery by Enrico Fermi, and members of his group, of what was later called the "moderator effect," the way to turn fast neutrons into slow ones, and the startling new phenomena those neutrons could induce. By itself, the discovery would be of interest only to fellow physicists, and ultimately to those good people in Stockholm. But about four years after Fermi's discovery, in a publication in which Lise Meitner and Otto Frisch immediately recognized the evidence for nuclear fission, Otto Hahn and Fritz Strassmann referred to the key role in their work of "slowed-down neutrons" (without happening to mention the Italians who had found how to make those slow neutrons). It can be said that on a day in October 1934, the clock began to tick which ever since has marked the nuclear age in world affairs.

Unintended Consequences

Despite their diversities, these examples, and others of this sort throughout history, have in common not only the initial unintention on the part of the discoverers but also the extraordinary transformations they eventually caused. They are of the rare sort of research findings that do not correspond to the more usual ones, being not merely discoveries of new facts, or verifications of predictions, or answers to old questions, or support to prop up an unstable theory. They are not just the addition of another brick to the ever-unfinished Temple of Isis. Rather, they suddenly open up access to an area beyond the map of established knowledge, thereby permitting an exploration of a new continent of *fruitful ignorance*. For what is most prized in science is the discovery of a vast absence of knowledge, of a range of hitherto undiscovered truths, owing to the breakdown of a standard model.

Superficially, those examples of profound discoveries might tend to support the view that the course of history itself is decided by the works of great men, to use the title of Wilhelm Ostwald's famous book. That opinion is contrary to the other old illusion, that it is history which shapes the ideas and acts of even the greats. But each of these two opinions is itself illusory. Any study of actual cases soon shows that both mechanisms together are constantly at work. As the psychologist Erik Erikson put it, "an individual life is the accidental coincidences of but one life cycle with but one segment of history." Even the most fateful chance observation has its own prehistory; conversely, there is no proof that even the most turbulent event in world affairs has been caused by some overarching *Zeitgeist*.

Preparing for a Nonexistent Career

From childhood on and into his early student years, young Fermi was recognized by his teachers, acquaintances, and friends as a prodigy.

Relying largely on self-study—a mode typical of great scientists, from Kepler to Faraday to Einstein—he soon became precociously at home with modern physics, enjoying equally the experimental and the theoretical sides. As a very young man Fermi turned to quantum theory, probably the first to do so in Italy, where that subject was considered a sort of no man's land between physics and mathematics rather than a promising research site. That part of physics was not taught in Italian universities, and a dissertation in theoretical physics as such would have been shocking.

Therefore, Fermi's dissertation at the University of Pisa (finished at age twenty-one) had to be an experimental one—on images obtained with monochromatic x-rays by means of a curved crystal. To build the necessary apparatus, Fermi persuaded his fellow students, Franco Rasetti and Nello Carrara, to help him—a first hint of his capacity for organizing teams. Typical also of his later years, Fermi was not satisfied with putting in print the experimental findings (in his seventh published paper, dated 1923). Before that he had published a separate, lengthy theoretical paper on the properties and theory of x-rays. There he showed that he commanded the whole literature—including von Laue, Bragg, Moseley, Barkla, Sommerfeld, Maurice de Broglie, Debye, Scherrer, etc.—in all their several languages. And already then he was keeping physics almost constantly in his thoughts. There is a story, perhaps apocryphal but believable, that one of Fermi's friends once found him pacing up and down in a room, with a preoccupied look. Concerned, his friend asked if Fermi was troubled by something. "No," Fermi replied. "I am just estimating by how much I am depressing the wooden floor as I walk along it."

Experimental x-ray studies, and even quantum physics, were by no means the only subjects enchanting the young physicist. It became important for his subsequent career that starting at age nineteen, Fermi's first five published papers were all on relativity theory. Most of them showed his mastery of the methods of general relativity, the theory just recently confirmed by Eddington's experiment. To be sure, most of the older generation of physicists in Italy were still skeptical and hostile to that theory. But like Wolfgang Pauli and Werner Heisenberg, at about the same time and at the same young age, Fermi had evidently been captivated by Hermann Weyl's new book, *Raum, Zeit, Materie,* for which Einstein himself had written an enthusiastic review in 1918. Fermi contributed to relativity a theorem of permanent value (later called Fermi coordinates), soon incorporated into textbooks on general relativity. Luckily, Italy had at that time several master mathematicians working in that field, such as Tullio Levi-Civita and Gregorio Ricci-Curbastro. They—and other mathematicians of first rank, including Guido Castelnuovo, Federico Enriques, and Vito Volterra—had begun to notice Fermi's papers and were ready to support his rise.

Yet, again contrary to ordinary expectations, Fermi properly soon realized that relativity theory was not the field in which to build his own career. From 1921 to 1925, he had no less than thirty-one publications, as reproduced in his *Collected Papers*, most in theoretical physics—even though he knew that there was not a single university chair available for it in all of Italy. His wide-ranging interests and sheer productivity were as astonishing as his self-confidence.

Since our theme is the formation of Fermi's team, I can point out that we have now already met the first member of the group that would soon be formed, the enormously talented experimental physicist and Fermi's schoolmate, Franco Rasetti. It is also time to introduce Orso Mario Corbino, a crucial figure in the eventual rise of Fermi and his group. Twenty-five years older than Fermi, Corbino was widely known for his early work in magneto-optics, for which he had been admired by Augusto Righi of Bologna, considered the previous generation's leading physicist in Italy. After Corbino had been called to the University of Rome, his talent as an administrator and unselfish connoisseur of talent quickly led to his becoming Senator of the Kingdom (1920), Minister of Public Instruction (1921), and Minister of National Economics (appointed in 1923, by Mussolini, although he never joined the Fascist Party).

Corbino's keen scientific mind, combined with his hope to put Italy again on the map as a center of great physics research, led him to mourn the sorry state of physics there, symbolized for him by Righi's death in 1920. He saw clearly that Italy was then unable to take advantage of the worldwide rise of opportunities in the new physics of the day. Without realizing it, by 1920 Corbino was ready to discover a Fermi—just as Enrico Fermi, for his part, must have known that without the help of such a man there might never be a Fermi group.

After Fermi's graduation from Pisa, he returned to Rome in 1922, living with his parents and his older sister, as he was to do for several more years—a member of a closely knit family. At the moment, he had neither a job nor prospects for one. But one day he decided to make a first visit to Corbino. The two men immediately took to each other. With Corbino's help, Fermi obtained fellowships, spending unhappy months at Göttingen and happy ones in Leyden under Paul Ehrenfest, then a couple of years in temporary posts at Florence, working with Rasetti. At last, in 1926, Fermi was appointed to the new Chair of Theoretical Physics at the University of Rome, a move engineered by Corbino—who was officially the director of the university's Physical Institute at Via Panisperna 89a, the building in which the top floor was in fact the flat of Corbino's family.

Establishing Patterns

Now Fermi could begin to put his and Corbino's dream into reality. But that was not going to be easy. So far, Fermi had admirers but no followers. The outlook for building a school of bright young physicists was still dark. There was not even an Italian text on atomic physics for advanced university students; and of those students there were only a handful, since the expectation for eventual university employment was extremely poor. Fast action was called for. First, Fermi wrote and published in 1927 that missing textbook. Corbino used his influence to bring Franco Rasetti from Florence to Rome, eventually settling him into a professorship for spectroscopy, created for that purpose. (How we all would have liked to have had a Corbino on our side! He seems the ideal candidate as patron saint for bright young scientists.) And now Fermi and Rasetti began to recruit promising university science students for their *Istituto*.

Emilio Segrè, a student in engineering in Rome, met Fermi, who was only four years older, with the result that Segrè knew instantly that here was an extraordinary teacher, scientist, and human being. That autumn, with Corbino smoothing the administrative problems, Segrè transferred his studies to the physics section of the university, thereby becoming Fermi's first pupil. In Segrè's words, "The Roman School had started."

Segrè in turn persuaded his friend Ettore Majorana to join the group, at least informally.

Here an important aside is called for. The previous paragraph contains several clues to the vitality and unique characteristics of the formation of the Roman School. First, Corbino was ever ready to help, in any way. Second, all of "Corbino's boys," as they came to be called later, were within a few years of the same age. Third, among them there was a camaraderie in which the only trace of hierarchy was the acknowledged centrality of Fermi's brilliance. Finally, almost all members of the group were part of one social network. They typically even spent parts of their vacations together at the seaside or in the mountains. For example, in the summer of 1925, Fermi was in the mountains with the families of Levi-Civita, Castelnuovo, and Ugo Amaldi. Amaldi's seventeen-year-old son, Edoardo, was fascinated by the scientific talk and ended up accompanying Fermi on a bicycle tour of the Dolomites. A bonding had begun there which, together with Corbino recruiting him from the engineering class, resulted two years later in Edoardo becoming part of Fermi's physics group at the institute.

If all this sounds a bit like the behavior of a stereotypical Italian family, let us remember that this was not the way things then generally arranged themselves in physics laboratories in, say, Göttingen or, for that matter, in New England. At any rate, we see that a critical mass was being formed in Rome. The group's younger students became more and more competent, partly through participating in experiments with Rasetti, but above all through Fermi's constant care and the value of his informal theoretical seminar. For Fermi was an ideal teacher—with one occasional exception: as John Marshall later recalled, Fermi sometimes saw to it that he was the only person near the blackboard who had the chalk, and it was very difficult to argue with the only person who had the chalk.

Fermi's typical mode of teaching was to keep things clear and seemingly improvised. He distrusted abstract theories such as the quasi-philosophical Copenhagen versions of quantum mechanics, favoring instead the visualizable approach of Schrödinger. Hans Bethe referred admiringly to Fermi's way as "enlightened simplicity."

Fermi thought and taught about physical phenomena in a unique way. Just as his experimental equipment functioned well despite its often being assembled from cannibalized pieces and in the least complex manner, so also did Fermi consider Nature herself put together in the most parsimonious way. That is to say, he recognized again and again the same pattern to be at work in completely different contexts. Thus he applied the same idea of scattering length first to explain the pressure shift of spectral lines (Document 95 in Fermi's *Collected Papers*), and second to understand artificial radioactivity produced by neutron bombardment (Document 107)—even using the same diagrams in the publications. Or again, he applied the same statistical theme to atoms on the one hand and to neutrons on the other. Fermi's great paper on beta decay, at its core, treats the emission of electrons and neutrinos in nuclear events as analogous to the emission of photons from atoms in excited states (Document 80b). As Fermi's colleague at the University of Chicago (and co-author on two papers), the astrophysicist Subrahmanyan Chandrasekhar, put it (Fermi, *Collected Papers*, vol. 2, p. 923): "The motions of interstellar clouds with magnetic lines of force threading through them reminded him of the vibrations of a crystal lattice; and the gravitational instability of a spiral arm of a galaxy suggested to him the instability of a plasma and led him to consider its stabilization by an axial magnetic field."

One recognizes here the way a thematic presupposition guides some scientists' understanding of how Nature works at the fundamental level. Einstein's basic assumption was again and again that entirely different phenomena are aspects of one grand unity. Niels Bohr, quoting a saying of Friedrich Schiller, thought that truths may be found best "in the abyss" between contrary theories. Fermi thought of each phenomenon as exhibiting one of only a relatively small number of different basic scenarios of which Nature availed herself; and of these, Fermi kept a catalogue throughout his life.

Preparing for an Opportunity

In order to learn new skills, members of Fermi's group, already international in outlook, visited laboratories abroad. Rasetti travelled to Millikan in Pasadena and later to Lise Meitner in Berlin. Segrè went to Pieter Zeeman in Amsterdam and Otto Stern in Hamburg. In turn, more students from other universities transferred to join the Rome group, including Eugenio Fubini, Ugo Fano, and Bruno Pontecorvo. They were attracted by Fermi's work, for example on the quantum theory of radiation, on statistics, above all on the theory of beta decay (the paper first published in 1933, after having been rejected by the editor of the journal *Nature* as "containing abstract speculations too remote from physical reality"). Also, a good number of young physicists came from abroad, to visit and sometimes to stay for longer periods and collaborate. They included Hans Bethe, George Placzek, Felix Bloch, Rudolf Peierls, Fritz London, Edward Teller, Eugene Feenberg. And before that, there were collaboration with and visits from colleagues at other Italian universities, such as Renato Einaudi from Turin, but perhaps most frequently from the newly flourishing physics group in Florence under Antonio Garbasso, including Bruno Rossi, Gilberto Bernardini, Giuseppe Occhialini, Enrico Perisco, Giulio Racah, and Sergio De Benedetti.

But what would be the freshly hatched young Roman group's contribution to physics? Up to 1929, the primary output of their teamwork was still in spectroscopy. But from then on, it became clear that remarkable changes in physics abroad required a new direction if Italian physics was to reach world-class level. The historical development of physics itself revealed to the Fermi team what these young men had been preparing themselves for, in all those years of wide-ranging study and perfection of various skills. The quantum mechanics of Bohr, Heisenberg, Pauli, Dirac, Schrödinger was taking center stage in the field of theory; and on the experimental side, nuclear physics was being transformed in exciting ways, by the findings of James Chadwick, Harold Urey, Clinton S. Davisson and Lester Germer, Carl Anderson, J. Curry Street and Seth Nedermeyer. The proton-neutron model of the nucleus was becoming plausible; the neutrino hypothesis of Pauli was tantalizing; and E. O. Lawrence's cyclotron was a much-envied sensation.

In a speech in September 1929, Corbino showed he had already sniffed out that *nuclear physics* was, in his words, "the true field for physics of the future." Now, the core members of the team, who had patiently and often at great personal cost stuck together for years, re-educated themselves in systematic study. Amaldi led a special seminar on radioactivity, and the group learned how to build neutron sources, construct a cloud chamber, and make Geiger counters. Some additional research funds were obtained from the Italian National Research Council (CNR). And all this was being done without the group realizing precisely what and when an opportunity would come along for using that new knowledge, to achieve the ultimate desire of the team: to make, at long last, a world-class discovery. It was a curious moment in the history of science: Arguably the first modern research team in physics, it was waiting for something to happen that could put the group to use for a high purpose.

A Field of Exciting Ignorance

Remarkably soon, something did happen, entirely out of the blue. In Paris, the French physicist Irène Curie Joliot and her husband Frédéric Joliot had been sending alpha articles from polonium into a cloud chamber to bombard aluminum, so causing the immediate emission of positrons from the target. One day, in January 1934, Joliot noticed by chance that the emission of positrons persisted when the polonium source was taken away. A new radioactive isotope had been created by the alpha particle bombardment. One might add here that a *non*-discovery of artificial radioactivity had earlier taken place in E. O. Lawrence's cyclotron laboratory in Berkeley. As Lawrence confessed in his Nobel Prize speech (for 1939, but given in 1951), "Looking back, it is remarkable that we [at Berkeley] managed to avoid the discovery," by neglecting the fact that the Geiger counters continued their chatter even after the 27-inch cyclotron had been turned off.

The Joliot-Curies' discovery opened a window on a new landscape of exciting ignorance. While others immediately rushed to explore this territory, using alpha particle sources, Fermi made the crucial decision to see if a beam of neutrons would also produce artificial radioactivity. It seemed to him reasonable to expect that the lack of charge of neutrons, many of them emerging at high energies from radon-beryllium sources, would permit a great effect on the charged nuclei of the targets, despite the admittedly still relative weakness of the sources available to them.

On March 25, 1934, Fermi was able to publish the first results of the group's experiments, in the journal of the National Research Council, *Ricerca Scientifica*. It was the first of ten such publications in the spring of 1934, some appearing just a week apart. From the third to the tenth of these publications, the list of authors was always given as follows: "E.

Amaldi, O[scar] D'Agostino [a young chemist who happened to come back to Rome on a vacation from a fellowship in Paris but happily was pressed into service], E. Fermi, F. Rasetti, E. Segrè." Note that all core members of the "family" were listed in alphabetical order and that, perhaps for the first time in the whole physics literature, there were as many as five accredited authors.

Fermi typically had decided to test all available chemical elements for artificial radioactivity, going methodically down the periodic table. The team divided the labor in a cooperative way—getting the targets, monitoring the electric circuitry of the Geiger counters, making chemical analyses, etc. Fermi's group, from the beginning, generally tended to work together on one project—unlike the operation at, say, Rutherford's own Cavendish Laboratory, where different small groups worked on different projects whose commonality was chiefly that they represented different parts of Rutherford's wide-ranging interests.

The work in Rome was now quite frantic and exhausting for some months, and a few mistakes were made. None was later regretted by Fermi's group more than the presumed identification of transuranium radioactive products, produced by irradiating the elements thorium and uranium with neutrons. The same mistake was made by others at the same time, including the Joliot-Curies. One is tempted to be thankful for that, because otherwise fission might have been discovered earlier, when only two countries might have been interested in putting it to use in a weapon—Germany and the Soviet Union. At any rate, altogether the Roman team irradiated over sixty elements with neutrons, producing by artificial radioactivity forty-four new isotopes.

Here was excellent work, the unexpected first fruits after the long wait and preparation. The Fermi group was now widely noted. Because at the time such publications had to be by law first in Italian, I. I. Rabi at Columbia University is said to have advised, "Well, now we all have to learn Italian." Rabi also began to consider the prospect of Fermi joining the physics faculty at Columbia—as was eventually to happen, with spectacular results.

The Two Tables

Meanwhile, without realizing they were now standing at the threshold of a startling discovery that was to have resonance long into science and world history, the whole Roman group took a break for the lengthy summer vacation, getting away from the heat of Rome. When they reassembled at the *Istituto* in the fall of 1934, they were joined by Bruno Pontecorvo, a close family friend of Rasetti. But by mid-October things began to go wrong. Their whole experimental activity was upset by a strange inconsistency in the results of irradiation of targets during an attempt to calibrate the degrees of induced radioactivity. The readings turned out to depend on the tables used as support of the equipment. One table, once the bearer of the spectroscopes of earlier days, was made of wood; the other, not far away, was a shelf made of stone. When an experiment on inducing radioactivity, in a target made of silver, was made on the wooden table, Pontecorvo noticed a markedly greater activity resulted than if the same experiment was tried on the stone support. The group christened it the "miracle of the two tables."

To get to the bottom of this puzzle, Fermi initiated a few days of systematic observation, starting October 18, 1934. He and his team reasoned that perhaps the lead housing around the target affected the neutrons reaching the target in those two cases, and they observed that the interposition of a block of lead changed the activation somewhat. Therefore, Fermi decided to make a lead filter, a wedge of varying thickness, to insert into the neutron beam. As Segrè put it later, on that day, "Persico and Bruno Rossi [were] there on a visit, kibitzing." The account of the events on that morning was later told by Fermi to Subrahmanyan Chandrasekhar, who published his report (Fermi, *Collected Papers*, vol. 2, p. 927). The essential last paragraph was repeated verbatim by others close to Fermi, such as Edoardo Amaldi and Emilio Segrè.[1]

> I will tell you how I came to make the discovery which I suppose is the most important one I have made. We were working very hard on the neutron-induced radioactivity, and the results we were obtaining made no sense. One day, as I came to the laboratory, it occurred to me that I should examine the effect of placing a piece of lead before the incident neutrons. And instead of my usual custom, I took great pains to have the piece of lead precisely machined. I was clearly dissatisfied with something: I tried every "excuse" to postpone putting the piece

1. A detailed version of the events was given by Laura Fermi in her book, *Atoms in the Family* (1954, p. 98). It differs in some details, but comes to the same conclusion.

of lead in its place. When finally, with some reluctance, I was going to put it in its place, I said to myself: "No! I do not want this piece of lead here; what I want is a piece of paraffin." It was just like that: with no advanced warning, no conscious, prior, reasoning. I immediately took some odd piece of paraffin I could put my hands on and placed it where the piece of lead was to have been. (Ibid.)

The result was immediately obvious: a great increase in the radioactivity induced in the target, even if the target and the paraffin filter were placed on the stone shelf. As Segrè recalled, at about noon, "everybody was summoned to watch the miraculous effect of the filtration by paraffin." And in a "still extremely puzzled" state, "we went home for lunch and our usual siesta." "When we came back at about three in the afternoon, Fermi had found the explanation of the strange behavior of filtered neutrons. He hypothesized that neutrons could be slowed down by elastic collisions, and in this way become more effective—an idea that was contrary to our expectation" (in Segrè, *Enrico Fermi, Physicist*, p. 80).

As Amaldi reported, it was only later that the so-called $1/v$ law was determined, i.e., that the capture cross-section (λ_c) was inversely proportional to the speed of the neutrons at low velocities. But on that day the miracle of the two tables was unmasked. Fermi realized that the hydrogen nuclei in the wooden table had greatly slowed some of the incident neutrons, being of about the same mass, and then had scattered them to the target, producing the unexpected effect on it, whereas the heavy nuclei in the stone table could slow and scatter neutrons only very poorly. Repeating the experiment quickly by using water instead of paraffin helped prove Fermi's initial hypothesis. Moreover, the enhanced radioactivity was also observed for copper, iodine, and aluminum.

That evening, in Amaldi's home, they all met to prepare a short report of their work for the *Ricerca Scientifica*, with Fermi dictating, Segrè writing, Rasetti, Amaldi, and Pontecorvo excitedly adding their comments. Amaldi's wife, Ginestra, who was working with that journal, saw to it that the article would be published within two weeks. Preprints—another novelty—were made available within days and sent out to some forty of the most prominent researchers in the field. Soon the whole profession knew that the Roman group had crossed a new frontier. It was a climactic moment for Fermi's team in Rome, and—as it turned out—for the world on its path into the uncertain future.

Estimation by Intuition

But in that story so far, there is a haunting puzzle. Fermi was the most rational and least impulsive of scientists; yet, not by accident or chance but by sudden, determined action, "with no advanced warning, no conscious, prior reasoning," he had placed the crucial piece of paraffin before the neutron source.[2] This seems to me an example of a kind of intuitive intelligence which sometimes guides brilliant minds in the early phases of their research. That concept is now rarely mentioned, least of all by scientists themselves, who tend to shy away from such difficult-to-define notions. But it had figured prominently in the writings of philosophers such as Baruch Spinoza, Immanuel Kant, and Henri Bergson. Arthur Schopenhauer, widely read at the time, even held that intuition is the hallmark of genius. Einstein referred to it as *Fingerspitzengefühl*, a feeling at the tips of one's fingers, and considered it essential to scientific inquiry (e.g., in his essay, *"Motive des Forschens,"* 1918): "There is no logical path to the elementary laws, but only intuition, resting on empathy gained by experience." Henri Poincaré (in *Science and Method*, Book II, chap. 2) noted that it is by logic that we prove but by intuition that we discover. The scientist and philosopher Michael Polanyi wrote at length about what he called the scientist's "tacit knowledge," largely resulting from one's lengthy immersion or "in-dwelling" in the subject of research. He summarized the result in the simple sentence: "We know more than we can tell." Much earlier, Hans Christian Oersted provided

---

2. One historian of science was so astonished by the report of an action so uncharacteristic of Fermi that he even doubted the account reported by Chandrasekhar. But that idea must surely be dismissed. Chandrasekhar, who was one of the most distinguished and precision-minded scientists, even felt it necessary to start his report with a footnote: "His [Fermi's] account made so great an impression on me that though this is written from memory, I believe that it is very nearly a true verbatim account." Moreover, he published it (in 1965) when those who had "kibitzed" that morning in 1934—Rossi and Persico—were still alive; and as mentioned, Fermi's co-workers, Segrè and Amaldi, endorsed Chandrasekhar's account as given above, quoting it in full.

for this rare gift the happy term "anticipatory consonance with nature." And a chemist and great writer, Primo Levi, wrote, "I know with my hands and my nose, with my senses" (in *The Voice of Memory*, p. 8).

So one might understand that when Fermi's hand was reaching for the "odd piece of paraffin" instead of the lead wedge, he was guided at that moment by a speculation below the level of consciousness, a result of an intimate knowledge of neutron physics, one built up during his years of intense study, discussions, and experimentation with neutronics. As Dr. Alberto De Gregorio[3] has shown in two articles, Fermi may well have read publications in 1932–33 in which slow neutrons and effects of neutrons on hydrogenous substances were discussed, and he also had participated in the 1933 Solvay Conference, which included discussions of these topics. But it is significant that nobody other than Fermi and his group entered into the crash program of producing artificial radioactivity, first with fast neutrons and then with slow ones, when Fermi was able to draw on resources that had slipped below the conscious level.

In fact, Chandrasekhar's account, given above, is part of a longer piece of his, which reveals that his whole discussion with Fermi had begun precisely with a consideration of the role of "subconscious" ideas in creative work in science: Chandrasekhar wrote (p. 926 in Fermi's *Collected Papers*):

> I described to Fermi [Jacques] Hadamard's thesis regarding the psychology of invention in mathematics, namely, how one must distinguish four different stages: a period of conscious effort, a period of "incubation" when various combinations are made in the subconscious mind, the moment of "revelation" when the "right combination" (made in the subconscious) emerges into the conscious, and finally the stage of further conscious effort. I then asked Fermi if the process of discovery in physics had any similarity. Fermi volunteered and said [there followed his account, as given above].

---

3. "Chance and Necessity in Fermi's Discovery of the Properties of Slow Neutrons," *Il Giornale di Fisica*, 42, no. 4 (2001): 195–208; and "Enrico Fermi e la scoperta degli effetti delle sostanze idrogenate sulla radioattività indotta dei neutroni," in P. Tucci, A. Garuccio, and M. Nigro, eds., *Atti del XXIII congresso nazionale di storia della fisica e dell'astronomia* (Bari: Progedit, 2004), pp. 154–164.

There are also other accounts of Fermi's ability to dredge up, from hidden resources, answers to questions facing him. Herbert Anderson recalled that at a crucial moment during the difficult early work in 1939 at Columbia University on the possibility of a chain reaction, under Fermi's direction, "Fermi asked to be left alone for 20 minutes," after which he emerged with a rough estimate of the effect of resonance absorption by uranium. Anderson reported that the estimate, which proved to be correct, "was largely intuitive. Fermi was never far wrong in such things." One can imagine the positive effect such talent had on Fermi's collaborators. Elsewhere Fermi was even credited with helping reactor engineers to obtain a rough estimate of data not yet measured, such as nuclear cross-section. They did it, reportedly, by watching Fermi closely for an "involuntary twinkle in his eyes" while reciting to him possible cross-section values. Indeed, the speech by Hans Pleijel of the Swedish Academy at the awarding of the Nobel Prize to Fermi, in December 1938, deployed that key word. Pleijel told his audience: "Along with Fermi's significant discoveries, and to a certain extent equivalent, can be placed his experimental skill, his brilliant inventiveness, and his intuition." Fermi's collaborators often said he worked "with 'CIF', an acronym for *con intuito formidabile.*"

Science historians have struggled to understand the mechanism behind such examples of "anticipatory consonance with nature." It seems to me another case of finding ourselves before an ocean of fruitful ignorance.

The Fall of the Roman School

Not long after its triumph, Fermi's group in Rome underwent a more and more rapid disintegration and near destruction—to use Edoardo Amaldi's phrase—experienced also by the research groups in Florence and elsewhere in Italy. In the early days of Fermi's ascent, some scientists had been helped at least indirectly—in hastening appointments and in the availability of some funds—by the Fascist government's attempt to revive the idea of Rome as the center of Western civilization. Among the sixteen institutions founded after Mussolini's assumption of dictatorship in 1925, two were the new Royal Academy of Italy and the National Council of Research (CNR). Fermi's work had required financial support from both, although he himself, by nature and design apolitical, felt repugnance for the regime's ideology. In turn, the government expressed annoyance with Fermi for his refusing the prestigious chair in physics left by Schrödinger in Zurich, where Fermi, acting as a proxy for Italy's science, would have been highly visible throughout the European continent. The state's police also kept a constant eye on Fermi, as the file still in the Italian State Archive shows.

From the mid-1930s on, a whole set of institutions founded by the Fascist government withered, part of the steady erosion of civilized life. The physicists at the institute in Rome submerged themselves in hard work, hoping to use physics as "soma," on the model of Aldous Huxley's victims in *Brave New World*. As their state lurched toward the Ethiopian campaign that started in October 1935, the economic and political dependence of Italy on Nazi Germany grew greatly. In July 1938 came the institution of racist laws in Italy, roughly along the German model. Most of the group's members emigrated. Fermi and his family left Italy in December 1938, bound first for Stockholm, then for the U.S.A. During World War II the authorities in Italy caused the arrest, deportation, and death of persons close to the Roman group, including Laura Fermi's father, Augusto Capon. Among the victims sent to the death camps were several relatives of Segrè.

So few were the years between Fermi's launching the spectacular recovery of Italy's place in physics in the mid-1920s and its descent by the end of the 1930s in horror and flames. Indeed, that arc of this story—the brave rise of Enrico Fermi and his group from bleak beginnings, their hard-won achievements, and then their ghastly dissolution—may well take its place among the symbols of the best and the worst in the history of the twentieth century.

References

Amaldo, Edoardo, *Twentieth-Century Physics, Essays and Recollections: A Selection of Historical Writings by Edoardo Amaldi*. Singapore: World Scientific, 1998.

Bernardini, Carlo, and Luisa Bonolis, eds. *Conoscere Fermi*. Bologna: Edizioni Scientifiche Società Italiana di Fisica, 2002. [English translation in press.]

Bethe, Hans, et al. Special Issue: Portraits of Fermi. *Physics Today,* 55, no. 6 (June 2002): 28–46.

Buck, Barbara. *Italian Physicists and Their Institutions, 1861–1911.* Ph.D. diss., Harvard University, 1980.

Fermi, Enrico. *Collected Papers,* vol. 1. Chicago and London: University of Chicago Press, 1962.

——— *Collected Papers,* vol. 2. Chicago and London: University of Chicago Press, 1965.

Fermi, Laura. *Atoms in the Family.* New York: American Institute of Physics, 1987.

Gambassi, Andrea. "Enrico Fermi at Pisa." *Physics in Perspective,* 5 (2003): 384–397.

Holton, Gerald. Chapter 5 in *The Scientific Imagination,* rev. ed. Cambridge, MA: Harvard University Press, 1998.

Pontecorvo, Bruno. *Fermi e la fisica moderna.* Rome: Editori Riuniti, 1972.

Proceedings of the International Conference: Enrico Fermi and the Universe of Physics. Rome: ENEA [Ente per le Nuove tecnologie, l'Energia e l'Ambiente], 2003.

Segrè, Emilio. *Enrico Fermi, Physicist.* Chicago and London: University of Chicago Press, 1970.

——— *A Mind Always in Motion: The Autobiography of Emilio Segrè.* Berkeley, CA: University of California Press, 1993.

Weiner, Charles, ed. *History of Twentieth-Century Physics.* New York: Academic Press, 1977. [See especially the essay by Eduardo Amaldi.]

— 6 —

B. F. Skinner, P. W. Bridgman, and the "Lost Years"

Some historians are encouraged by a saying of Jean le Rond d'Alembert, who noted that our task is to comprehend "our forefathers to whom we owe everything and about whom we know nothing." Having known the psychologist B. F. Skinner and the physicist-philosopher P. W. Bridgman, and having witnessed the complex relationship between them, I wish to share some thoughts on these major scientists to whom their respective fields are in many ways indebted. Their relationship, which initially seemed destined to be warm and fruitful, finally was unrequited, and yet of importance to both. It also may have been one of the missed opportunities in the history of ideas.

Skinner and "the Lost Years"

For Skinner's side of the story, I need not go deeply into the biographical details.[1] Entering Hamilton College in 1922, Skinner, somewhat aimlessly, discovered his interest in writing and literature, getting his B. A. in English literature in 1926. But reading Bertrand Russell and John B. Watson, he absorbed the idea of science as a means for the reconstruction of society. That is of course an ancient and recurring hope. It was repeated, for example, in the famous 1929 pamphlet of the Vienna Circle, entitled *"Wissenschaftliche Weltauffassung: der Wiener Kreis,"* issued as part of the "Ernst-Mach-Verein," under the authorship of Rudolf Carnap, Hans Hahn, and Otto Neurath. It was this very message which helped decide W. V. Quine, one of Skinner's closest intellectual companions at Harvard from the earliest days, to go to Vienna to participate in the activities of the Vienna Circle and then on to Prague to study under Carnap.

Entering Harvard for graduate study in 1928, Skinner initially still seemed to have no clear direction. In this searching spirit, he sat in on a course on the history of science under the charismatic Lawrence J. Henderson, professor of biological chemistry and head of the Harvard Business School Fatigue Laboratory, and George Sarton, the father of the modern history of science. I believe it was Henderson who suggested in 1929 that Skinner read Ernst Mach's *Science of Mechanics*. In that book he encountered Mach's missionizing arguments against metaphysics and for a coherent world picture and the economy of thought.

The book had a permanent effect on Skinner. In an interview on June 8, 1988, Skinner stated to me categorically: "I was totally influenced by Mach *via* George Sarton's course, and quickly bought Mach's books, *Science of Mechanics,* and *Knowledge and Error.*" As E. A. Vargas has pointed out, Skinner did not stop there: writing in 1931 to W. J. Crozier, he exulted "I have also been reading Mach's *The Principles of Physical Optics,* which is one grand book. The best summer reading I have come across."[2]

Skinner was ready to receive Mach's empiricist message. As he stated in his autobiography, *The Shaping of a Behaviorist* (1979), he recalled only two science books he had read as an undergraduate: Jacques Loeb's *Comparative Physiology of the Brain and Comparative Psychology* and *The Organism as a Whole,* with their largely positivistic approach to the study of the behavior of animals. When Skinner came to Harvard University to do his graduate work in 1928, his thesis supervisor, in whose laboratory he remained for five years, was the physiologist W. J. Crozier. It is not accidental that Crozier's own teacher had been Jacques Loeb, who in turn had been an admirer and correspondent of Mach. Indeed, "it was the ultra-positivistic form of Loebian biology that Skinner encountered at Harvard."[3]

1. What follows draws on well-known sources, including of course Skinner's own books; on L. D. Smith and W. R. Woodward, eds., *B. F. Skinner and Behaviorism in American Culture* (London: Associated University Presses, 1996), and in particular the essays in it by Nils Wiklander, "From Hamilton College to Walden Two: An Inquiry into B. F. Skinner's Early Social Philosophy," and John J. Cerullo, "Skinner at Harvard: Intellectual or Mandarin?"

2. E. A. Vargas, "Prologue, Perspectives, and Prospects of Behaviorology," *Behaviorology,* 3, no. 1 (1994): 112. I have written on the effect of Mach on Skinner in Chap. 1 of my book *Science and Anti-Science* (Cambridge: Harvard University Press, 1993).

3. Smith and Woodward, op. cit., p. 277.

Skinner's doctoral dissertation at Harvard, dated December 1930, was entitled "The Concept of the Reflex in the Description of Behavior." The introductory pages read as if they really had been written under the influence of Mach; for they warned, as Mach had done in his analysis of physical concepts, particularly those of Newtonian mechanics, that much of the field of the description of the behavior of intact organisms is beholden to "historic definition, that is to say, was vested with extrinsic interpretations, and some of these now appear to embarrass the extension of total behavior." In fact, on the second page, Skinner shows his source:

> The reader will recognize a method of criticism first formulated in respect of scientific concepts by Ernst Mach and perhaps better stated by Henri Poincaré. To the work of these men, and to Bridgman's excellent application of the method to more modern concepts [in P. W. Bridgman's recently published *Logic of Modern Physics*, 1927], the reader is referred for an extended discussion of the method *qua* method. Probably the chief advantage first exploited in this respect by Mach lies in the use of an historical approach. . . . [T]he second part of the thesis [which is "primarily experimental"] is thus offered as an example of the practicability of the method, and as a partial test of the hypothesis, advocated in the first part.

As if to make the point quite clear, on page 55, the section "Notes and References for the First Part" begins with just five books: Ernst Mach's *Mechanics* (1883), Mach's *Analysis of Sensations* (1886), Henri Poincaré's *Science and Hypothesis* (1903) and *Science and Method* (1908), and, last but not least, Bridgman's *Logic of Modern Physics*—a reinforcement of Skinner's earlier mention of Bridgman's "excellent application of the method."

At this point a puzzle suggests itself. Bridgman, well known to be an admirer of Mach and in other ways congenial to the kind of epistemological approach Skinner was favoring, was the only one of the three authors mentioned who was still alive, indeed at the height of his powers—and right there at Harvard. It would have been eminently reasonable for Skinner, who did not lack courage, to have sought out Bridgman, eighteen years his senior but kindly and accessible. It could have been the start of a productive interaction, but apparently it did not happen. Here was perhaps the first of a series of lost opportunities.

In writing his doctoral thesis, young Skinner saw a way of applying the Machian point of view to the clarification of such concepts as the "reflex" of intact organisms, something he considered to be as basic in psychology as, say, mass is in physics. As Skinner recollected, he was "following a strictly Machian line, in which behavior was analyzed as a subject matter in its own right as a function of environmental variables *without reference to either mind or the nervous system*"; that was "the line that Jacques Loeb . . . had taken."[4] In this radically empiricist mode, the study of behavior was reduced, for Skinner, to the observation of the motion of the foot of a food-deprived rat pressing down a small lever in an experimental box of standard size. Explanation was reduced to description, causation to the notion of function, and the chief goal was the correlation of observed events. Again, Bridgman would have been sympathetic to this train of thought.

Skinner's Ph.D. was awarded in 1931. Thanks to receiving a National Research Council fellowship and his election to the elite Society of Fellows at Harvard, he had five years, to 1936, of postdoctoral freedom. He began a manuscript that was to be his first scholarly book, "Sketch for an Epistemology" (of which only a 1935 article was published, on the generic nature of stimulus and response). The project showed, however, that, as Skinner often said, epistemology was his first love. In a way now almost totally foreign to contemporary scientists, epistemology and history of science were part and parcel of his work, and undergirded it with enormous strength. His reading in philosophy, science, and literature was extremely wide (he kept a log of what he was reading, which not surprisingly included works not only by Loeb and W. B. Cannon but also by Einstein). While Skinner was busy with finding "quantitative regularities and orderliness in the behavior of rats," it was *in the context of what science could say to the rest of culture,* as we shall see at more length in a moment. I find it not surprising that in the same year of 1931, when he developed the final form of the lever-press box that became his key to the development of the concept of the operant, he also became a member of the fledgling History of Science Society.

By 1938, Skinner left Harvard for an instructorship at the University of Minnesota, then briefly was chair of the Psychology Department at

---

4. B. F. Skinner, review of Smith's *Behaviorism and Logical Positivism,* in *Journal of the History of the Behavioral Sciences,* 23 (1978): 204–209, on p. 209 (emphasis in original).

Indiana University. Again, if fate had played him a different card, and if Skinner had stayed on in Cambridge just to the beginning of September 1939, he would undoubtedly have been caught up in a unique and congenial event, the International Congress for the Unity of Science—itself an heir of Mach and his followers—held at Harvard on September 3–9, 1939, just as war broke out in Europe. Bridgman, with the physicist-philosopher Philipp Frank, was the main organizer, and Quine served as secretary of the meeting. Henderson, Sarton, S. S. Stevens, Carnap, and Kurt Lewin were among the dozens of speakers presenting papers. Skinner would have felt very much at home there. But equally important was that of this Congress Quine could later write simply, "Basically this was the Vienna Circle with accretions, in international exile." In fact, this occasion was the beginning of a series of monthly meetings of like-minded colleagues across all fields in the Cambridge area, which convened most vigorously in the 1940s. The spirit of this inter-science discussion group was again characterized by Quine as "a sort of revival of the Vienna Circle," headed by Bridgman and Frank.[5]

The discussions of the group ranged very widely over topics of importance to intellectuals of the time, but above all they were a forum for striving mercilessly for clarity and commonalties. A typical example is noted in an announcement for January 8, 1945: "Professor Richard von Mieses will lead a discussion on 'Sense and Nonsense in Modern Statistics.'" The next meeting's invited speaker was Charles Morris from New York, who was to make more clear what scientists say when they "seek simplicity or economy in their theoretical work." Talcott Parsons spoke on "Psychoanalysis and the Theory of Social Systems"; Norbert Wiener on "The Brain and the Computing Machine," in which he gave the first draft of cybernetics; George Wald on "Biology and Social Behavior;" John Edsall on "The Life and Work of Walter Cannon." Others who led evening discussions were Wassily Leontief, Hudson Hoagland, I. A.

---

5. For detailed descriptions of these meetings and the movement behind it, see Chap. 1, "Ernst Mach and the Fortunes of Positivism," in my book *Science and Anti-Science,* and in my two articles expanding on the subject: "From the Vienna Circle to Harvard Square: The Americanization of a European World Conception," in F. Stadler, ed., *Scientific Philosophy: Origins and Developments* (Dordrecht: Kluwer, 1993), pp. 47–73, and "On the Vienna Circle in Exile: An Eyewitness Report," in E. Köhler, W. Schimanovich, and F. Stadler, eds., *The Foundational Debate* (Dordrecht: Kluwer, 1995), pp. 269–292.

Richards (long a friend of the absent Skinner), Jeffries Wyman, Crozier, Roman Jakobson, Quine, Edwin Boring, Henry A. Murray, John von Neumann, Oscar Morgenstern, Howard Aiken (on the first electro-mechanical computer), and Bridgman (on the problem of meaning).

The energy and excitement in many of these discussions were enormous. Being much the youngest of the group, I was persuaded without much resistance to be the secretary of this movement—in charge of keeping track of the meetings, circulating the attendance sheets (many of which have survived), and writing up summaries. One can only speculate, of course, how Skinner's thought and career would have been influenced by these meetings, as was the case for many of the participants—including myself.[6] Skinner's absence was thus again one of those "lost opportunities" of which intellectual history abounds.

In 1947, Skinner was back at Harvard to give the William James Lectures. He joined the Harvard faculty in 1948 as a professor, at the instigation chiefly of Boring and Stevens. Skinner's 1938 book, *The Behavior of Organisms*, had begun to bring him the recognition that he had been waiting for. His research flourished, chiefly in the style he had discovered during the writing of his thesis, but a new aspect with respect to his general vision of himself became more and more prominent in his work. Unlike most other scientists, who are content to stay at their lab bench, he saw a function for himself in the wider society. In his study of Skinner at Harvard, John J. Cerullo[7] properly draws attention to the fact that it would be wrong to think of Skinner as a mandarin in the original sense: an honored and specially selected intellectual who could be counted on to support the existing societal structure, as was the case in the Chinese imperial period. On the contrary, Skinner was more nearly of the type Karl Mannheim defined as the "free-floating intellectual," somewhat of a rebel with respect to contemporary social structures or interests but at the same time, in the classic sense, a *Kulturträger*, who saw his function not only to uphold but also to improve the contemporary cultural-social order.

Today this spirit is largely languishing among the intellectual classes,

---

6. Among other participants in these meetings were Karl Deutsch, Philipp Frank, Edwin C. Kemble, Hans Margenau and Ernest Nagel (as visitors), Talcott Parsons, Harlow Shapley, Laszlo Tisza, and Paul Samuelson.

7. See footnote 1.

in which there are few civil intellectuals in this sense, and not only among social scientists. As Cerullo put it (p. 217), the job of our universities is now to turn out, at best, more mandarins. In contrast, Skinner's "advocacy of scientific method, biological determinism, and social control" (p. 217) was a calling that often put his career at risk in the early stages; it certainly went far beyond what was expected of the average experimental psychologist. As Skinner himself wrote to Edward M. Freeman in March 1937:[8] "A society ultimately depends on its top crust of intellectuals. If nationalism prevails, human society will have adopted the principle of the anthill and the beehive. Intelligence must protest and can hardly fail to triumph." Eliminating "mentalisms" was for him only a precondition to becoming an intellectual leader; his goal to do what the work-a-day psychologist—and, as we shall see, even Bridgman —would have regarded as hubris, namely to "make behaviorism a scientific force" in national life and "himself a leader of it" (Cerullo, p. 222). Cerullo adds, "Behaviorism was his 'higher and wider truth,' a culture-regenerating—indeed world-transforming—mission to which other activities were properly subordinated" (p. 230).

Skinner's book *Walden II*, for example, cannot be understood properly without seeing the author's intention to search for a radically new base upon which to build the culture of the future—even as many of the original Vienna Circle members also did in their own way. In no other manner could one explain Skinner's two pungent chapters in the middle of *Beyond Freedom and Dignity* (first published in 1971), entitled, "The Evolution of a Culture" and "The Design of a Culture"—not to mention his swipe at the "pure" scientists, "in the sense of being out of reach of immediate reinforcers" (p. 166).

After Skinner's return to Harvard in 1948, he was at last put in touch with the inter-science discussion group I mentioned before. By that time it had been in operation for about a decade—even though its energy was beginning to dissipate or diffuse. Skinner took active part in a public conference "On the Validation of Scientific Theories" in December 1953, chiefly organized by the discussion group. On that occasion

---

8. Letter in the Harvard Archives' Skinner Collection; still incomplete and being supplemented, the collection now contains 30 feet of letters and manuscripts.

he presented a "Critique of Psychoanalytic Concepts and Theories" (a topic that has become ever more relevant).[9]

P. W. Bridgman

At that conference, with dozens of speakers from near and far away—arguably both the high point and the ending of the operations of that movement in Cambridge—Bridgman was of course also present and active. Let me turn now to the other side of this story, to set the stage by giving some background about Bridgman—a colleague of mine, as was Skinner, but one whom I knew longer and better.[10]

I believe that Bridgman could not be fully understood unless one watched him day after day, while he was doing his lab work. He came in early, usually on his bicycle, before everyone else except the shop crew. Lean and relatively small of stature, in splendid physical condition until his last years, he would change immediately into his well-used lab coat, work on new equipment at the lathe (he made much of his apparatus himself), and race back and forth between the equipment and measuring instruments during his daily "runs." Except for his part-time helper in the shop and an assistant, also on a part-time basis, who helped him with measurement readings, he preferred to be alone in his fairly narrow and cramped surroundings. Exuding energy and seriousness, he accomplished in a few hours a run that might take a day or two for a dissertation student like me, working nearby. Watching him, I could not help thinking of a passage from Goethe's tragedy, *Faust*, in which the scholar opens up the first page of the Bible, encounters the phrase "In the beginning was the Word," and energetically rebuts "In the beginning was the Deed." I was seeing operationalism hyper-active before my eyes.

---

9. The Proceedings of the whole conference were published in the *Scientific Monthly* in 1954 and 1955 and then republished as a book by Beacon Press in 1956 (*The Validation of Scientific Theories*, edited with an introduction by Philipp G. Frank). A glance at the book will show how difficult it would be to bring together today such a high-powered set of thoughtful speakers on such a variety of aspects of the culture of our time.

10. I have written on Bridgman as scientist and philosopher on a number of occasions, most recently in the eleventh chapter, entitled "Percy W. Bridgman, Physicist and Philosopher," in the first edition of my book *Einstein, History, and Other Passions* (Woodbury, NY: American Institute of Physics Press, 1995).

Bridgman accepted very few graduate students to do their Ph.D. theses under his direction during his long membership on the faculty at Harvard (from 1908 to 1954). Avoiding almost all university committees, he dedicated himself to research and writing, reflected in the high and steady output of papers, chiefly on physics but also on philosophy of science. Bridgman published on the average about six substantial papers a year, with such titles as "The Resistance of 72 Elements, Alloys, and Compounds to 100,000 Kilograms per Square Centimeter." His lifetime total was over 260 papers, in addition to over a dozen books. Most of his writing was remarkably personal, often in the first-person singular. He was essentially the father of high-pressure physics as a field of research, and it was for this that he was awarded the 1946 Nobel Prize in physics.

Starting with a maximum obtainable pressure of 6,500 atmospheres, through his own design for sealing the pressure vessels and other ingenious inventions, with utter concentration and skill, and despite a remarkably low annual budget, he drove the experimental range of pressures up to an estimated 400,000 atmospheres toward the end of his active research. Along the way he discovered spectacular new behavior of matter. The materials under study were of course enclosed in massive steel cylinders, with a couple of long and very thin holes running down the length of the cylinder, the only access to the samples within being the wires monitoring the sample. On that point he once wrote, in his typically matter-of-fact way, "It is easy, if all precautions are observed, to drill a hole . . . 17 inches long in from seven to eight hours"—in carbon-alloy steel.

Plato said that "clear ideas drive away fantastic ideas." For Bridgman, the need to clarify his ideas in honest self-evaluation was not merely therapeutic, it was a biological necessity. He had an ethos of lucidity and candor of the most difficult kind: with himself. In 1938, doing better work than ever, he wrote, "As I grow older, a note of intellectual dissatisfaction becomes an increasingly insistent overtone in my life. I am becoming more and more conscious that my life will not stand intellectual scrutiny, and at the same time my desire to lead an intelligently well-ordered life grows to an almost physical intensity."

In these words we can identify one of the important themes in his life. Bridgman's struggle was always with himself, with his own understanding, with his desire to think situations through to his own satisfaction. The word *operation*, he explained later,[11] was first explicitly used in a discussion at a 1923 meeting of the American Association for the Advancement of Science on relativity theory, involving G. D. Birkhoff, Harlow Shapley, and himself. His *Logic* was, as it were, a summary of what he had gone through to alleviate his "intellectual distress" regarding relativity theory. The book was written during a half-year sabbatical leave in 1926, in great haste. He was always aware of its incompleteness, even though the whole book gained from the vast outpouring of energy and, if I may use the word, spirit that infused it. One result of this book was that he was under constant external and internal pressure to elaborate his notions, and he had to watch others, often to his dismay, "erect some sort of a philosophic system" on his work.

Bridgman never was or could be a disciple of any "ism," including operationalism, or positivism, or behaviorism. He always tried to eliminate metaphysics, as Mach had done, to throw the spotlight on performable action, above all an action performed by himself. Ultimately, he was a private man, so much so that he was accused of solipsism, to which he scarcely objected. He gave no university courses on his views, which others called a philosophy, although he wrote extensively on them. Even when his point of view was adopted in various versions in other fields, ranging from psychology to economics, he did not applaud, but rather watched with bemusement or dismay.

In keeping with his general attitude, Bridgman was not impressed with those who tried to squeeze his operationalism into a formalism favored by logicians, and he especially refused the concept that others—such as the "scientific community" as a whole—could decide on scientific truth. The existence of those "others," so important in the philosophy of science of empiricism and to sociologically informed views of how scientific truths are generated by consensus, counted for him very little. Since he truly had conceptual difficulty with the idea of those "others," his ever-skeptical mind was ready to question what to me were sometimes the most obvious points. For example, after we had become colleagues in the department and had established a pleasant relation-

---

11. P. W. Bridgman, "The Present State of Operationalism," in *The Validation of Scientific Theories*, pp. 75–76.

ship, he gave me the manuscript of his book, to be entitled *The Way Things Are,* and asked for my comments. However, he added, "I really wanted to call it, 'How It Is,' but the publisher didn't like that. The truth is I am not so sure that 'things' *are.*"[12]

As he put it in his essay "Science: Public or Private," in *Reflections of a Physicist* (1950, p. 56):

> The process that I want to call scientific is a process that involves the continual apprehension of meaning, the constant appraisal of significance, accompanied by a running act of checking to be sure that I am doing what I want to do, and of judging correctness or incorrectness. This checking and judging and accepting that together constitute understanding are done by me, and can be done for me by no one else. They are as private as my toothache, and without them science is dead.

While there was no trace of "relativism" in his research in physics, one reason for his intensely personal understanding of the epistemology of science was that he realized that formalisms may require for their full meaning the personal background and understanding of the individual. This includes unexamined assumptions that may or may not be adequate, not to speak of what Einstein called, to the consternation of positivists, the necessity of a quasi-intuitive *Fingerspitzengefühl* for the subject of scientific inquiry.

Without having an appreciation of Bridgman's internal psychological strength and confidence, one cannot understand how this essentially lonely man could do so much magnificent scientific work.[13] He said, "I stand alone in the universe with only the intellectual tools I have with me. I often try to do things with these tools of which they are incapable,

---

12. Here, Bridgman again indicated his sympathy with the Vienna Circle philosophy; Otto Neurath had urged explicitly the avoidance of "metaphysically" laden words such as *reality* and *things* (in the *International Encyclopedia of Unified Science* [Chicago: University of Chicago Press, 1938–39], vol. 2, no. 1).

13. Bridgman's sense of isolation comes through in a letter of March 30, 1938, to Philipp Frank, whom Bridgman expected to see with "great pleasure": "My work is done practically alone. I have no students [in philosophy of science] and have practically no contact with members of the department of philosophy and, in fact, most of them are not at all sympathetic with our point of view. The only young philosopher here whom I have particularly interested is Dr. Quine."

and I have often been misinformed and have delusions as to what they are capable of; but nevertheless it is my concern and mine only that I get an answer."[14] He could of course not neglect the stream of commentary, both favorable and unfavorable, launched by his book and articles on philosophy of science. He famously had to add "paper and pencil operations" to the more physical or "instrumental" ones, to accommodate, among other things, mathematics. He constantly had to fight against the idea that there is a "normative aspect to 'operationalism' [or 'operationism,' two terms that he said he 'abhorred']," which is understood as the dogma that definitions *should* be formulated in terms of operations. "An operational analysis is always possible, that is, an analysis into what was done or what happened. . . . [It can be given] of the most obscure metaphysical definition such as Newton's definition of absolute time. . . . It must be remembered that the operational point of view suggested itself from observation of physicists in action."[15]

On some occasions Bridgman found himself in the position of reassessing his own writings publicly. One was in 1959, when I asked Bridgman to contribute a self-analysis and review, after three decades, of his own book, *The Logic of Modern Physics*, to be published in the quarterly *Daedalus*. As one might expect, it was a thoroughgoing and rather negative critique of all the things that he now thought he should have added, changed, or understood differently in the writing of the book. In reading his self-analysis we begin to see even more clearly the growing distance between himself and Skinner.

Bridgman said there that he was particularly embarrassed to have written in *Logic*, "We should now make it our business to understand so thoroughly the character of our permanent relations to nature that another change in our attitude such as that due to Einstein shall be forever impossible. It was perhaps excusable that a revolution in our mental attitude should occur once, because after all physics is a young science, and physicists have been very busy, but it would certainly be a reproach if such a revolution should ever prove necessary again" (p. 520).

---

14. P. W. Bridgman, "New Vistas for Intelligence," in *Reflections*, p. 370. In the end he pronounced that famous definition, "The scientific method consists of doing one's damnedest with one's mind, no holds barred" (pp. 57–58).

15. *Validation of Scientific Theories*, p. 79.

To this he now added in retrospect:

> To me, now it seems incomprehensible that I should ever have thought it within my powers, or within the powers of the human race for that matter, to analyze so thoroughly the functioning of our thinking apparatus that I could confidently expect to exhaust the subject and eliminate the possibility of a bright new idea against which I would be defenseless. [After all] *our skulls contain a simply appalling number of undiscovered structures which must condition and limit our thinking. . . . In fact, it seems to me that for us here and now the problem of adequately understanding the nature of our minds and what we can do with them is a problem more pressing, and perhaps more difficult, than the problem of understanding the physical world.* . . . The comparatively new methods of the brain physiologist and the behavioral psychologist are without doubt of the greatest value and should be pushed as aggressively as possible. *But the older methods, methods of "introspection," if you like, are not to be discarded,*[16]

Bridgman was now ready to emphasize more the "mental" operations (p. 522).

Skinner vs. Bridgman

To Skinner, all this must have verged on unacceptable "mentalism." To a behaviorist, the mind and consciousness are inaccessible to true scientific study. Moreover, psychologists such as Stevens had become ardent operationalists, but Stevens too had moved far from Bridgman's position when he wrote that "An essential characteristic of all facts admitted to the body of scientific knowledge is that they are public. Science demands public rather than private facts. . . . Scientific knowledge has what we may call a social aspect."[17] In return, Bridgman wrote in the following year[18] to a colleague about Stevens that he admired much

---

16. P. W. Bridgman, "*The Logic of Modern Physics* After Thirty Years," *Daedalus*, Summer 1959, pp. 520–521 [emphasis supplied].

17. S. S. Stevens, "The Operational Basis of Psychology," *American Journal of Psychology*, 47 (April 1935): 323.

18. Letter of May 4, 1936, quoted in Maila L. Walter, *Science and Cultural Crisis* (Palo Alto, CA: Stanford University Press, 1990), p. 184.

about him, "but I simply cannot make him see that his 'public science' and 'other one' stuff are just plain twisted. I have also discussed with him his 'basic act of discrimination' without making much impression, and I have rather washed my hands of him."

Indeed, even before Skinner had returned to Harvard, the seeds had been planted for a prolonged debate between him and Bridgman, the man whose work had been one of his main guides during his thesis-writing years. In "The Operational Analysis of Psychological Terms,"[19] Skinner—who we must remember had early in his career and to some degree throughout it been enchanted with literature, writing, and language—wrote that he "consider[ed] the language descriptive of private experience as the product of social reinforcement. Accordingly these private events are inferences supported by 'appropriate reinforcement based upon public accompaniments and consequences,' and being conscious, which is nothing more than 'a form of reacting to one's own behavior, is a social product.'" But Bridgman, whose own skull housed a brain that he believed to be rather impermeable to social reinforcements and products, protested, and his rejoinder was quite sharp: "In the private mode I feel my inviolable isolation from my fellows and may say, 'my thoughts are my own and I'll be damned if I let you know what I am thinking about.'"[20] Even earlier, Bridgman felt upset by the need to deal with Skinner's objections, writing (Bridgman Archives at Harvard, November 7, 1953), "I shall probably continually have Skinner in the back of my head, imagining that I am discussing or arguing with him."

Bridgman's crusade was in large part intended to rescue the individual from what he saw as the collectivization of society all around him. The unique individual and the uniqueness of the individual—very American ideas—always remained close to his heart. Certainly, as a member of society, he would act in a "public mode" as a good citizen. But what counted most was that secret inner volcano which, inaccessible to all others, was his true preoccupation, day and night. In resignation, Skinner wrote to Bridgman: "My efforts to convince you of the

---

19. *Psychological Review*, 52 (September 1945): 270–277.

20. The quotations are from Maila L. Walter, *Science and Cultural Crisis*, pp. 188–191. They refer to Bridgman's article "Rejoinders and Second Thoughts," *Psychological Review*, 52 (September 1945): 281–283.

possibility of extending the operational method to human behavior have long since suffered extinction."[21]

There is finally another reason why these two genial men were ultimately not more attracted to each other intellectually: the difference in their ultimate agenda, beyond doing good science. This comes out eloquently in one of the few letters from Skinner contained in the Bridgman archives. Shortly after his return to Harvard, Skinner decided to give a freshman course, Psychology 7, entitled simply "Human Behavior." The catalogue description was brief but ambitious: "A critical review of the theories of human behavior underlying current philosophies of government, education, religion, art, and therapy, and a general survey of relevant scientific knowledge, with emphasis upon the practical prediction and control of behavior."

That description is a simple indicator of a profound difference between our two protagonists. Whereas Bridgman was chiefly interested in getting things straight in his own head, Skinner could be said to have had the active agenda of both a researcher *and* a culture-improver, even in his undergraduate courses. On June 13, 1949, Skinner wrote to Bridgman:[22]

> I thought you might be interested to look over the enclosed notes which summarize my lectures in Psychology 7 this past term. The course will be given better another year but I feel reasonably satisfied that I have impressed some of my students with the implications of a scientific attack upon human behavior. I am very serious about this course and although it is cutting into my research time I hope to go on giving it. I don't know of anything more important at the moment than to acquaint the lawyers, politicians, diplomats, business men, journalists and educators of the future with the methods of science.

Skinner added, in a gesture indicating a willingness for cooperation and conciliation: "I hope to make better use of your own writing along this line when the course is given again."

Apparently Skinner and Bridgman met at lunch in June 1958 to discuss their differences face to face. Although they remained friendly col-

21. Letter of May 10, 1956, quoted in Walter, p. 192, letter in Bridgman's papers.
22. Harvard Archives, Skinner Papers, HUG(FP)60.10, Box 1.

leagues in other matters, however, the essential gulf remained. When Skinner first came to Harvard, there was potential for harmony, perhaps even a collaboration, between these two seminal minds. They had a common intellectual parentage, a shared scientific/philosophical culture. Perhaps a bridging between them was never to be, but they might have been drawn closer if they had developed a companionship instead of being separated during those exciting "lost years."

The main issue between them was, after all, reminiscent of the division in physics between the world of the palpably macroscopic and the untouchably submicroscopic. In that case, a posture of complementarity developed: each of the two opposing sides was right within its domain, but they also fit together in a larger context. I still feel that in the field of ideas on which these two protagonists battled, an eventual solution of the same sort will be found. Working together, Skinner and Bridgman might have been the first to find that breakthrough.

— 7 —
I. I. Rabi as Educator and Science Warrior

I. I. Rabi was first of all a superb physicist; the head of a most productive laboratory at Columbia University that attracted brilliant students and postdocs; then Associate Director of the MIT Radiation Laboratory; Senior Advisor to Robert Oppenheimer at Los Alamos; after World War II, again a physicist and consultant to government, laboratories, and international organizations. Throughout his career he was a most respected senior statesman of science—and always the voice of reason, of moral authority, sometimes able to hide his commitment behind his down-to-earth humor. Less well known are two additional facets of Rabi's life, apparently separate from his scientific and political pursuits: Rabi as eloquent spokesman for an education system of wide scope and highest standards, and Rabi as a passionate defender of the view that a modern culture must prominently include science. He even had the courage to give one of his books the title: *Science: The Center of Culture*.[1]

It is on those two aspects that I shall concentrate in this essay, using preferably his own words to indicate his cast of mind. I also hope to show that Rabi's interests in education and in the place of science in culture were not two different enthusiasms, set apart from his views on physics and statecraft, but rather were organically part of one coherent vision he held about his mission on this earth.

---

1. I. I. Rabi, *Science: The Center of Culture* (New York and Cleveland: World Publishing Co., 1970).

A Passion for Education

I caught my first glimpse of a source of Rabi's immense and effective energy when I was chatting with him one day about a physicist who had been one of his students but who later enjoyed being a rather rough customer. With an impish glance, Rabi said: "Well, you must remember that there are two kinds of physicists: one kind turned to physics because in early life they had trouble with their radio kits; the other became scientists because they had trouble with their God. Our friend . . . well, he is of the first kind."

I didn't understand until much later how serious Rabi's remark was. At the time I took his reference to the second type to refer to some persons in the history of science who sought truth first in religion and then in science. The most famous case of course was Albert Einstein (see Chapter 1). I will return to Rabi's remark shortly.

A year or two later I had another talk with Rabi. I had been asked by the National Science Foundation in the post-Sputnik early 1960s to develop a national physics course for high schools, one meant not only for future scientists but for all students. That was how the Project Physics Course started, in which we integrated physics with some of the history and the methods of science to show that the evolution of physics was part of the rise of modern Western civilization, and also that even the grand masters of our science initially had trouble developing their counterintuitive ideas.

With the intention of asking Rabi to head the advisory committee of the project, I described its intended approach. To my surprise he agreed immediately. The project seemed to touch him as filling a need he was already deeply convinced of. He accepted my request, and moreover allowed us to print, at the top of the first page of the text, the following impromptu remarks as I recorded them:

"Science is an adventure of the whole human race to learn to live in and perhaps to love the universe in which they are. To be part of it is to understand, to understand oneself, to begin to feel that there is a capacity within man far beyond what he felt he had of an infinite extension of human possibilities. . . .

"I propose that science be taught at whatever level, from the lowest to the highest, in the humanistic way. It should be taught with a certain historical understanding, with a certain philosophical understanding, with a social understanding and a human understanding, in the sense of the biography, the nature of the people who make this construction, the triumphs, the trials, the tribulations."[2]

I should have remembered that much earlier, in 1955, Rabi entitled his Morris Loeb lecture at Harvard "Science and the Humanities." There, foreseeing the rise and costs of what is now called the "science war," Rabi spoke up for the blending, in his words, of "these two traditions," science and the humanities, "in the minds of individual men and women." This was two years before C. P. Snow's famous lecture on the intolerable gap between the "two cultures." Indeed, Snow later admitted privately that while Rabi was in London in 1957, Rabi was "the man who gave me the idea for the two cultures." The source of this information is John Rigden's fine biography, *Rabi, Scientist and Citizen,* to which I shall be referring often in what follows.[3]

Another eye-opening source I found while investigating Rabi's views more fully was the interview he gave in 1963 to the Quantum Physics Oral History Project, a unique joint activity of the American Physical Society and the American Philosophical Society. The first question the interviewer asked was how Rabi initially got into quantum physics. Rabi's reply may well have surprised his interrogator: "I was raised in a very Orthodox Jewish family with a great religious influence on me, and it is my view, perhaps, that my scientific interest came from the religious ... [especially] the first chapter of Genesis. ... On the other hand, my favorite reading and my best subject in high school was a subject in which I got very high grades very, very easily and without any work at all was history; history was in that sense my top subject, although my

2. *Project Physics Course* text (New York: Holt, Rinehart and Winston, 1970 and subsequent editions), Preface.

3. John S. Rigden, *Rabi: Scientist and Citizen* (New York: Basic Books, 1987), reissued with a new preface (Cambridge, MA: Harvard University Press, 2000). See also Lloyd Motz, ed., "A Festschrift for I. I. Rabi," *Transactions of the New York Academy of Sciences* (1977); Norman Ramsey's biography of I. I. Rabi in the *National Academy of Sciences Biographical Memoirs,* 62 (1993): 310–325; Norman Ramsey, *Spectroscopy with Coherent Radiation* (River Edge, NJ: World Scientific, 1998); Rabi's Oersted Lecture, delivered before the American Association of Physics Teachers and published as "Nature Revealed: The Joys and Dangers of Experimental Physics," *American Journal of Physics,* 50 (1982): 972; and Rabi's essay in I. I. Rabi, ed., *Oppenheimer* (New York: Scribner, 1969).

interest was in science which, I think, came about from Genesis. . . . [Also] my first reading of the Copernican theory of the explanation of the seasons and so forth left me with a scientific interest which never flagged, and I can still contemplate the Copernican system with a tremendous amount of pleasure."[4]

What a mixture of Genesis, history, and science! But it will make sense. Rabi was the first-born child and only son of an impoverished family of Orthodox Jews, who brought him from a small town in Galicia soon after his birth in 1898 to America, first to the lower East Side of Manhattan. Going to Hebrew school from age three, he quickly learned to read, of course above all the Bible, and he always delighted in the Bible stories, which he knew by heart.

The practices of strict Orthodox Judaism, from childhood on, include praying to God throughout the day, from opening one's eyes in the morning, on first washing, and into the night. Rabi stopped following these rituals probably by his teens. Eventually he took a position similar to Einstein's concept of Cosmic Religion, turning against the institutionalized forms of religion, which he thought tend to divide people and produce conflict. But something important remained from his upbringing. Rabi once reported: "There is no question that basically, somewhere way down, I am an Orthodox Jew. . . . To this very day, if you ask for my religion, I say 'Orthodox Hebrew'—in the sense that the church I am *not* attending is *that* one. . . . It doesn't mean I am something else."[5] "My early upbringing, so struck by God, the Maker of the world, this has stayed with me."[6] In his own way, Rabi remained God-struck throughout his life. He once reminisced: "[Physics] filled me with awe, put me in touch with a sense of original causes. Physics brought me closer to God. That feeling stayed with me throughout my years in science. Whenever one of my students came to me with a scien-

---

4. American Physical Society and the American Philosophical Society, *Quantum Physics Oral History Project*. Interview with I. I. Rabi, December 8, 1963. Courtesy of the Archives for the History of Quantum Physics, Center for History of Physics, American Institute of Physics, College Park, MD.

5. Jeremy Bernstein, chap. 2, "Rabi: The Modern Age," in *Exploring Science* (New York: Basic Books, 1978), p. 46.

6. Rigden, op. cit., p. 21.

tific project, I asked only one question, 'Will it bring you nearer to God?'"[7]

In my research on scientists, I try to be alert to the resonance between their cultural backgrounds and their later achievements. In Rabi's case, the resonance is clearly present. By the time Rabi was a teenager he had developed a mindset which guided him throughout his life, a worldview which looked for transcendent and elegant order and coherence, whether it was triggered initially by the Bible or by the Copernican system. At age thirteen or fourteen, having become aware of the disparities and inequities in the social world, Rabi even toyed briefly with Marxism. He later said of this phase: "What Marxism gives you is a view of society and a view of history—an *integrated* view. It's mostly wrong, but it is a view. You get the habit from this of thinking of things in a holistic way. You see connections.... I had the advantage of a religious background. Religion is also a system that encompasses everything, but it has something that Marxism doesn't have: religion has color and class. That whole idea of God, that's real class."[8]

Against this background, let us now ask in what sense Rabi was an *educator*. Young Rabi, without resources or mentors, had the drive and unbounded curiosity to become a self-educator—an old theme among major scientists in their youth, from Faraday to Einstein. Rabi would typically read four or five books a week. He was especially fascinated by history, and in fact could report that at high school he got "the highest grade in the state on the New York Regent's history exam."[9]

Similarly, after enrolling at Cornell University as a chemistry major, he was talented enough not to have to labor over his courses and so could follow his widespread interests on his own. He reported later, "So if I was taking a course in qualitative analysis, I might be reading Freud. I did an immense amount of reading. In this strange way, without taking courses, I got myself a liberal education. I would read about the history of the subject I was taking, the history of science.... [Mine was a] more relaxed approach, the historical, more synoptic view; I got more

---

7. Ibid., p. 73.
8. Ibid., p. 26.
9. Ibid., p. 29.

pleasure out of the thing than the others did, and I didn't have to do as much work."[10] To Lee DuBridge he said once, while reminiscing about his undergraduate education, "If you decide you do not have to get A's, you can learn an enormous amount in college."[11]

So Rabi as educator was first of all his own educator, driven by his wide-ranging desire to understand. That helped him also to realize that the part of chemistry he liked most was called physics. What drew him to physics was, not surprisingly, again the unifying power of a few great ideas and laws.

Finally, on entering graduate studies at Columbia University at age twenty-five, he discovered the new physics coming from Europe, not yet through courses at Columbia but through reading European journals of physics. Typically, again, Rabi organized a self-education group of fellow students, including Francis Bitter, Ralph Kronig, S. C. Wang, and Mark Zimanski. Rabi commented on this later: "I organized [this] group to study modern physics, that the faculty didn't teach.... I was getting around to things I had done as a little boy, when I organized a group."[12] These study groups became marathon sessions lasting from 11 A.M. into the late evening, for example, discussing the papers in the latest edition of the *Zeitschrift für Physik*. But Rabi also continued his routine of reading on his own, including the historical classics of science. In this way, he could later say that "One day I happened to be reading, for sheer pleasure, Maxwell's *Treatise* [of 1873]"[13] in the library. This work gave him the clue for quickly measuring the magnetic susceptibility of a crystal, which was central to his Ph.D. thesis project.

Similarly, when at the end of 1926 Rabi was stuck trying to apply Schrödinger's new version of quantum mechanics to molecular systems, he was rescued while reading for pleasure a book by K. G. J. Jacobi, the nineteenth-century German mathematician. There he found the equation Rabi had been unable to solve, with Jacobi's solution in terms of a hypergeometric series.[14] So it becomes clear that Rabi was not just try-

10. Ibid., p. 32.
11. *A Tribute to Professor I. I. Rabi on the Occasion of His Retirement from Columbia University* (New York: Department of Physics, Columbia University, June 1970), p. 13.
12. Rigden, op. cit., p. 42.
13. Ibid., p. 43.
14. Ibid., p. 52.

ing to say fashionable things when in his later pronouncements he insisted that history and a broad-based approach should have their place in good science teaching.

In 1927, a new theme enters our account. Like so many American scientists in the 1920s, Rabi felt that, after getting his doctorate, he had to go to Germany to learn at first hand the new quantum mechanics and experimental skills, above all to acquire the taste and style of doing physics on a grand scale. As he put it, he already knew the lyrics but he had to learn the melody. Leaving for Europe in July 1927 on a travelling fellowship, he met with and observed many of the most productive scientists of the time—Bohr, Born, Stern, Heisenberg, Pauli, Sommerfeld, Dirac, Schrödinger. From those, as he put it later, he learned what physics should be: taste, insight, standards to guide research, "a feeling for what is good."

In addition to finding, in Otto Stern's laboratory, the first steps to his own scientific excellence, Rabi came to a realization of historic consequence. As was generally agreed, America lacked at the time people at the top level in physics. But Rabi also noticed shrewdly that, compared with the German physicists in the rank just below those world-class scientists of Europe, he and many of the young American physicists then studying in Europe were as good as any of them. This opened his eyes to his next mission. As he put it later, "What we needed were the leaders."[15] In addition, during his two years in Europe, Rabi grew increasingly vexed by the general contempt he encountered toward American physics. Typically, it led him into action. Other Americans had also been irked by derogations of their work, including Ed Condon and Bob Robertson. Rabi called on them and they made a compact that "they would put an end to the second-class status of American physics."[16] When they returned to America they would not only do physics of importance and in the proper style but would undertake to become leaders of the field.

This decision was perhaps one of Rabi's most propitious ones as an educator in science in America. It released an immense amount of energy that not only propelled American laboratories to superb work but later also took on operational meaning in Rabi's interest in better sci-

---

15. Ibid., p. 63.
16. Ibid.

ence education at all levels. Rabi added: "We said, 'We were going to really do something, very definitely in our generation,' and we did, because ten years later the *Physical Review* was the leading journal in the world.... When we came back, my generation, Condon, Robertson, Oppenheimer, myself... people like Van [Vleck].... We just sort of changed the whole works in this short time. I had half the graduate students in the department at the time I was an assistant professor [at Columbia]. We sort of brought back the white man's magic, so to speak.... The general democracy which we had allowed young people to come forward."[17]

As the only son in an Orthodox family, Rabi had learned early that he could, as he said, "intimidate" his parents. Now, on returning to New York, he did it to the Columbia Physics Department: "I was all over the place as soon as I came back [from Europe]; you see, that's the kind of person I am.... Suddenly the whole damned department started working hard whereas they hadn't before. They said 'Goddamn it, Rabi, before you came here we used to have three hours for lunch and play bridge!'... I think I made life interesting from the scientific point of view... ideas for experiments, meanings, how paradoxes are explained, what quantum theory means in this situation or that situation. People just got very interested.

"We worked every day of the year with the exception of three or four days, New Year's day and Christmas day, and so on.... Once I got going on those experiments, this was a day and night job, all the time. I had about fifteen students to keep going."[18]

With students and postdocs of the caliber of Norman Ramsey, Frank Press, Leon Lederman, Jerrold Zacharias, and many others, Rabi was ideally positioned to inspire and set standards for a whole new generation of top-flight scientists. To be sure, his molecular beam research, into which he had thrown himself passionately, left little time for preparing lectures on quantum mechanics. Throughout the 1930s many of his students found his classroom lectures confused and disorganized, but they learned from his improvised groping at the blackboard and from the excitement he exuded, which made them rush to the books

---

17. *Quantum Physics Oral History Project* interview, op. cit., p. 29.
18. Ibid., pp. 35–38.

and to private discussions. Along the way they learned how to gain confidence.

Frank Press said about a course he took from Rabi: "If he made a mistake, a very simple mistake . . . the next thirty minutes or so would be spent trying to find that mistake. However, in terms of giving perspective to the significance of a particular discovery, the state of the art, the humanistic aspects of science, a sense of excitement—he was without peer."[19] And in any case, Rabi's door was always open after class for discussion. It may have been said of him at the time that he was "simply an awful lecturer"; but he also was a great teacher, and in that respect fitted into the charismatic chain from Kepler to Helmholtz and Rutherford, who all shared that characteristic.

Rabi was particularly effective in conveying to his students and co-workers a sense of experimental elegance and the criteria for choosing a research subject. In addition, he had a long-range effect as an educator, through his influence on the work of other physicists in a great variety of fields, as well as through the application of his ideas, such as the magnetic resonance method, not only in physics but also in chemistry, biology, medicine, and so forth. Beyond that, his contributions ranged from maser research to the theory of quantum mechanics. He may not have liked to get involved in an experiment with apparatus he had not built, but he was the guiding force on the concept of the experiment, the equipment, the interpretation of results, and above all the standards and the choice of problems.

A Warrior for Science

With the end of World War II, Rabi took on an additional role. His political self-education during the Los Alamos period had convinced him that "the world had changed, and the future would be something which would not take care of itself," but would require some "practical politics."[20] With the authority that came with his Nobel Prize of 1944 and the public understanding of the role physicists had served in helping to rescue Western civilization from its enemies, Rabi became a public per-

---

19. Ibid., p. 71.
20. Rabi interview, ibid., p. 71.

son, a civic scientist. This came to him naturally, for he felt that intellectuals should live a life beyond their research specialty and, in addition, that the scientists of the day now had a particular ethical responsibility. As he put it, now that technological advances "have made it absurdly easy to kill human beings," the "disinterested search for objective truth" requires, because of the end products to which such a search might lead, also the shouldering of social responsibility. Significantly, these last five words appeared in one of his first essays for a wider public, in the October 1945 issue of the *Atlantic Monthly*.

From then on he would speak up again and again, not least on politics and ethics; for example, he expressed moral revulsion against President Truman's decision to develop the H-bomb and against the persecution of Oppenheimer. Others have written about his influence on the politics of the time; about his prescient and vigorous action for a European joint center for science, which became CERN; about his role in initiating the Atoms for Peace conferences held in Geneva in the 1950s, which brought together a massive confluence of scientists of all ages from both sides of the Iron Curtain. He became one of the most effective statesmen of science that America has had.

But I want to conclude with one aspect of his public life. He had a passionate commitment—which originated in the holistic, synoptic point of view acquired in his early youth—to safeguarding a rightful place for science as a necessary and liberating aspect of the total culture of our time. In lecture after lecture, Rabi was clear and specific about why this had to be so. Let me summarize his main points.

(1) Science can act as a *unifying force for all humanity*, and thereby it counteracts the divisive and destructive tendencies of almost all other human enterprises: "Science is the greatest, uncentralized, undirected cooperative effort of all time . . . among people of the most diverse origins and cultures."[21] It brings people of the most various backgrounds and with loyalties to entirely different social systems into a common search for truths.

(2) Science can act as a personally *ennobling activity*. As he said in the quotation we used in the *Project Physics* text, "Science is an adventure of the whole human race to learn to live and perhaps love the universe in

---

21. Quoted in Rigden, op. cit., p. 260.

which they are. To be a part of it is to understand, to understand oneself, to begin to feel that there is a capacity within man far beyond what he felt he had, of an infinite extension of human possibilities."[22]

(3) Scientists have to *defend themselves* against enemies who fail to see science as part of one unified culture. As Rabi put it in 1964, science is a counterforce to "the anti-Galileans."[23] In that lecture and in another three years later, Rabi took on the counterculture and anti-science movements, then already in full force. He had lived through two-thirds of a century marked by the excesses of irrationality on a vast scale. Unlike so many other scientists who are too preoccupied or timid to speak out on this subject, he called the "battle rag[ing] on every campus," the war against science, "a symptom of both ignorance and of a certain anti-rational attitude which has been the curse of our century.... It is a sort of poison which undermines the self-confidence that leads to the highest development of a rich culture."[24] In yet another lecture, he asked for dedication to build "a unified culture in which the sciences describe a world which is alive with people and with feelings, and the humanities describe a world in which the physical universe is not inert matter but rather is a part of the development of the human spirit."[25]

(4) Science *teaches how to think objectively, rationally, and therefore productively* for solving the real-world problems in society, and especially in a democracy. Speaking at Yale in 1962 on "Science and the Liberating Arts," he proposed that science "is the first of the liberating arts and perhaps the model for the others . . . [because it] will teach us to look at our problems objectively and solve them in the manner best suited to our needs and possibilities." "Scientific truth and the scientific adventure can set the standard for our contemporary striving for a just and meaningful world."[26] "After all, there is hardly a problem of government that does not have an important scientific aspect."[27]

(5) Finally, science can be a *good preparation for a useful life outside science:* "If science were taught more humanistically in the schools and

22. *Project Physics,* op. cit., Preface.
23. Quoted in Rigden, op. cit., p. 113.
24. Ibid., p. 32.
25. Ibid., p. 42.
26. Ibid., pp. 45–47.
27. Ibid., p. 90.

in the universities, it could become a foundation for any career, not only teaching or research."[28] That would give "vigor to our civilization." Rabi asked "Where does [one now] look for leadership, support, and implementation?" "Unfortunately, [one looks] to a Congress and Executive. . . . They are often ignorant if not definitely hostile to the finest flower of our culture, suspicious of intellect and reason, addicted to words without content or context. No wonder we are in trouble."[29]

He resented, he said as early as 1946, that "to the politician, the scientist is like a trained monkey who goes up to the coconut tree to bring down choice coconuts."[30] He added that intelligent people well educated in science and the humanities, including "artists, writers, economists, sociologists," should "enter into political life" and run for public office, to supplement what he saw as the "present incumbents, the lawyers and businessmen."[31] But Rabi was by no means guilty of scientism or in favor of a takeover by scientists. With his balanced judgment, Rabi agreed that President Eisenhower was correct in his Farewell Address to warn against the "danger that public policy could itself become the captive of a scientific technological elite."[32]

I conclude with this thought: Throughout Rabi's dazzling career, that self-educated, God-struck youngster within him was constantly motivated to wise and courageous service on behalf both of science and of the country he loved. I think he had before his eyes a model. In a lecture in 1964,[33] he confessed: "My ideal man is Benjamin Franklin—the figure in American history most worthy of emulation. . . . Franklin is my ideal of a whole man." Initially a craftsman, Franklin was one of the greatest scientists of his time while also standing out as diplomat and statesman. To this Rabi added poignantly, "Where are the life-size—or even the pint-size Benjamin Franklins of today?" From what we know of Rabi's words and deeds, it is clear that Rabi himself is a worthy, life-size model to aspire to in our own turbulent time.

28. Ibid., p. 92.
29. Ibid., p. 48.
30. Ibid., pp. 139–140.
31. Ibid., p. 49.
32. Ibid., p. 81.
33. Ibid., pp. 111–112.

— II —

SCIENCE IN CONTEXT

8

Paul Tillich, Albert Einstein, and the Quest for the Ultimate

My aim in this chapter is to share with you a glimpse I was privileged to get of the landscape of two great minds, Tillich and Einstein. In different but parallel ways, they both reached out to the limits of human understanding and were driven by what Tillich called Ultimate Concerns. Their ambitions were so enormous that, in the end, neither fully succeeded in attaining them. Yet, each left us an invaluable legacy, and perhaps also a lesson for the great challenges of today, so poorly attended to. These two men had much in common and might have become close comrades, but because they looked at the world from different perspectives, they came eventually into conflict.

I must first explain that my credentials for discussing Tillich do not rest on any claim that I am an intimate student of his theology, or that I am able to clarify some puzzles in his writings. Other scholars, including Mr. William Crout, who has kindly shared with me his collection of sources, are more familiar with Tillich's work. This body of work is extensive, including a vast archive in the Theological Library at Harvard's Divinity School, the official *Collected Works* in fourteen volumes, in addition to many other books and the more than 1,200 entries in the catalog of the Harvard Libraries that refer to works by or about him.

What I *can* claim is that I enjoyed Tillich's generous intellectual company during the nearly seven years while we were faculty colleagues at Harvard. We had many discussions, and appeared together at invited presentations on science and religion. Hannah and Paulus were frequent guests in Nina's and my house, and we in theirs. He accepted my invitation to be a consulting editor of the journal *Daedalus*, and in its first volume, for the year 1958, he let me publish his essay entitled "The Religious Symbol."

At this point I think I am expected to mention how I first met Tillich. It was a revealing meeting. In my case this encounter, in 1955, had been prepared by Harvard's president, Nathan Pusey. Mr. Pusey had the good habit of inviting, to a fine dinner, a dozen or more of his professors, those of us who were giving the various, large introductory courses in the General Education program of those years. On each of those occasions, some topic of general interest to this group's common task was discussed, in a spirit of gentle amity.

One evening, however, this amity came to an abrupt end. One of the professors there remarked in passing that he doubted the Divinity School might have anything valuable to contribute to the College's program. Mr. Pusey became visibly upset. Since his arrival as president about two years before, the Divinity School had been one of his main preoccupations. "Just wait," he said. "At one of our next meetings you will see an outstanding theologian, who has been attracted to our Divinity School."

Not long after, the new star joined us. It was Paul Tillich. Even during the casual dinner conversation before his talk, one sensed his special quality, his membership in the great European tradition of culture, his familiarity with high-level intellectual controversies, and also his liveliness at age sixty-nine. As his student and assistant, Paul Lee, said later, Paul Tillich was, "as a scholar, a one-man theological symphony," yet also, unlike some academics, "passionate, full of desire, vibrant with vitality." Indeed, one was immediately drawn to him.

Tillich had just come to Harvard from the Union Theological Seminary in New York, where he had found refuge after being forced to leave Germany. Dismissed from his professorship at the University of Frankfurt in 1933, Tillich had been an early and outspoken opponent of the Nazi regime.

In 1951 he had published the first volume of his master work, *Systematic Theology*, and was working on the other two volumes, issued in 1958 and 1963. Just before coming to Harvard he had given lectures at the University of Virginia on "Biblical Religion and the Search for Ultimate Reality." There he had said: "The God who is *a* Being is transcended by the God who is Being itself, the ground and abyss of every Being. And the God who is *a* person is transcended by the God who is the Person-Itself, the ground and abyss of every person. . . . *Against*

Pascal I say: 'The God of Abraham, Isaac and Jacob and the God of the philosophers is the same God. He is a person, and the negation of himself as a person."

I have selected these passages for two reasons: to prepare us for the conflict over the concept of the personal God, which will become important later in this account, and to alert you to the dialectical style in Tillich's thinking, a style which James Luther Adams, arguably his most eminent commentator, called "a philosophy of paradox."

To return to the dinner: When Tillich's presentation began, he turned to us with a question—Would we like him to outline two of his central ideas, which were, as he put it, "God is the infinite ground of Being," and "Mankind's highest duty is to focus on Ultimate Concerns"?

There was a long, awkward silence.

What might have troubled my colleagues in their silence? Surely, no problem with the word *God*. It was the first thing you expected to hear from a theologian. And in any case, even the scientists there knew that some of their kind had been practically on first-name terms with God. A famous example was the great experimental physicist, I. I. Rabi. He wrote once, as noted in Chapter 7, "[Physics] filled me with awe, put me in touch with the sense of original causes, brought me closer to God... Whenever one of my students came to me with a scientific project, I asked only one question: 'Will it bring you nearer to God'?." For his part, Einstein had said memorably: "What really interests me is whether God had any choice in the creation of the world." And to this day, Stephen Hawking, among others, is implicating the deity in his research findings.

So, if the assembly was not puzzling over the word *God*, perhaps it might be pondering over the word *Being*, in the phrase "God is the infinite ground of Being." That word in English is such a pale reflection of the complex and historically fought-over concept *Sein* or *Seiendes*. Tillich had been using these words since his earliest days in Germany, not only because he had had to study books like Hegel's *Die Lehre von Sein* or because for a time he had been a faculty colleague of Martin Heidegger (in Marburg).

Or perhaps Tillich's audience was trying to adjust mentally to the phrase "ultimate concerns," one of his favorite terms [*ultimate* in his German was *das Unendliche, das Unbedingte*]. But each of us there should have thought immediately of victors and victims in pursuit of their ultimate concerns, their grand challenges. Among scientists, history records that many of the best were driven to depression and even suicide. And of course, history, literature, song, and myth are full of those ecstatic or inconsolable seekers of the ultimate: all those desperate lovers, for whom the words *yes* or *no* are the "whole world" (think of young Werther, or Schubert's Wanderer); explorers at the extremes of the known world; or the world's great prophets and their ascetic religious devotees. To turn to the demonic side, on which Tillich also wrote much, think of the vast number of humans on our unhappy globe for whom the quest for the ultimate has spiraled down to the frantic search for survival for another day. Howard Nemorov, in a short work entitled *The Questor Hero*, traced the lives of the luckier ones, the heroes in the Holy Grail romances, calling them "the seekers, the questers, who range heaven and hell. . . . [The Questor] is forever searching for the grail—that is to say, the Highest: knowledge, wisdom, consecration. . . ."

Well, when the awkward silence at the table was broken at last, it turned out that what had confused some of Tillich's audience was another matter entirely: One of the puzzled scientists asked Tillich if he would care to define in what sense he was using the word *infinite*. Was it perhaps in the sense of the actual infinite of the great mathematician Georg Cantor, in his theory of transfinite sets? Or in the sense closer to other theorists, from Aristotle's Potential Infinity to our day?

Many around the table must have felt embarrassed—but not Paul Tillich. A smile came over his face: "Dear Colleagues," he said, "This is exactly why I have been looking forward so much to join you at Harvard. In my last position and in others before, I was surrounded mostly by theologians; so I had no opportunity to learn from them what I can learn from you. I would love to do that now." He had us in the palm of his hand.

A few of us there, including myself, gladly agreed to set up informal meetings with Tillich to learn from him and from one another. To preview one thing we quickly learned from Tillich: He told us that he thought of the meaning of ultimates and especially of infinity by means of visual metaphors. Visualization and symbolism in art were important to him. He routinely included pictures in his lectures, in order to show, as he put it once, "the possibility of breaking the surface of reality in order to dig into its depth . . . and you cannot understand theology without understanding symbols."

Exploring the symbolic use of an image, he said, would help us to understand how he perceived the "symbol" of infinity. Thus Tillich said that the idea of sitting at the edge of an ocean and gazing out was for him a symbol of the infinite: it hinted at an infinite depth before him, but it also bordered on the finite, as he was positioned on the beach, the boundary between the two. Here I remembered Friedrich Schiller's famous couplet, "Only fullness leads to clarity, and truth lies in the abyss" *(Nur die Fülle führt zur Klarheit, und im Abgrund wohnt die Wahrheit)*.

Tillich and His Mission

Let us look now more closely at some of Tillich's key ideas and how they originated, and then compare these with Einstein's own main motivating concept. For this purpose it would of course be interesting to know whether and when our two protagonists actually met. They might have done so early, in Berlin. Einstein was in Berlin from 1914 until 1932. Tillich came there in 1919, a time of great chaos, and until 1924 was Privatdozent at the university. He lectured on "A Theory of Culture," on a vast range of fields—politics, art, philosophy, psychology, and sociology. His work could be seen as a grand, multidimensional mission, specifically in the service of relating religion to the rest of culture and—being both pastor and social democrat—to real life.

While in a way they were colleagues, whether Einstein and Tillich actually met in person then in Germany remains a rumor. They can be said to have met then in another way important to our story, however: namely, by sharing much of the current worldview and cultural background as well as being, each of them, possessed by essentially the same ultimate goal, as we shall see shortly.

Because our own cultural barometer has, during the past decades, swung from unity to diversity, we may be puzzled over how the passion for generalization, for synthesis, came to each of our two protagonists. But Einstein, Tillich, and intellectuals of their generation—those born in the late decades of the nineteenth century—were exposed, in their schooling, reading, and discussions, to similar forces during their cultural formation. In their impressionable years they would each have read the standard classical authors, and above all one figure whom both Einstein and Tillich frequently referred to and quoted in their correspondence: Immanuel Kant.

One of the lessons many of Kant's nineteenth-century followers took from his *Metaphysical Foundations of Natural Science* was that two opposing forces determined all natural phenomena, but that this polarity only masked a "hidden [*versteckte*] identity." This idea allowed Kant to hypothesize the existence of a *Grundkraft*, one fundamental force of which all other forces are variants. It is of course a thematic line that goes back to antiquity, to Thales the Ionian, who looked for one substance or essence to explain all phenomena of the material world. A version of the Ionian Enchantment possessed Kant, who put unity first among his categories.

To those who regarded themselves as his pupils, Immanuel Kant provided the wellsprings from which issued two main directions of thought. One is exemplified in the scientific work of major nineteenth-century scientists such as Hermann von Helmholtz, Emil Du Bois-Reymond, and Rudolf Virchow. On the other side, Kant could be read, or misread, as the father of a very different view of science, one infused with the Romanticism of the "Nature Philosophers." They included Friedrich Schelling; the brothers Schlegel; Novalis; and all their influential followers.

The Danish *Naturphilosoph* Hans Christian Oersted had even proved experimentally in 1820 that the existence of a fundamental force was plausible. For Oersted showed that an electric current produces around itself a magnetic field. That was the first step in the synthesis of different fields, which was expanded by Maxwell in the 1870s to include light, by Hertz to verify for radio waves, and finally by Einstein, to include all of these and more. Parallel unifications were developed by others, such as the law of conservation of energy applicable to all sciences. The password for the sciences at the time was "Holism." And outside the sciences, the nostalgia for unity and synthesis was also pursued.

Tillich explained how he became infected with the synthesizing passion in an essay entitled "Autobiographical Reflections." It is as remarkable story. He passed his earliest years in East Germany, in small towns built around the Gothic church, all within the medieval town walls. His father was an authoritarian Lutheran minister and his mother, as he wrote, was also morally rigid. Young Tillich found refuge in what he called a "romantic," "aesthetic meditative attitude toward nature," reinforced by the "deeply moving" nature mysticism in beloved German poetry—Goethe, Hölderlin, Novalis, Nietzsche, George, Rilke. Reading poetry led him to a vision, as he put it, of the presence of the infinite in the finite, which he regarded as also theologically affirmed.

Eventually there were confrontations with his stern father, the "angry" supporter of "the conservative point of view." Tillich rebelled against this outlook both philosophically and politically, eventually becoming a prominent supporter of Germany's Religious Socialism. "The two strong motives I received," he wrote, were "the romantic and the revolutionary. The balance of these two motivations has remained the basic problem of my thought and of my life ever since."

In his Gymnasium years, he adored ancient Greek culture, especially the pre-Socratic philosophers Heraclitus and Parmenides, of whom he wrote later (in *The Future of Religion*, 1966) that nothing in all of philosophy written since then has surpassed them.

At that point in his autobiography, Tillich suddenly writes: "The way to synthesis was my own way. It followed the classical German philosophers from Kant to Hegel, and remained a driving force in all my theological work." On his own, starting when still in high school, young Tillich studied works of Kant, Fichte, Schleiermacher, Hegel, and Schelling. When he came to do his doctoral dissertation, it was on Schelling's philosophy of religion. Later he turned to the period when Schelling broke with Hegel's "system of reconciliation" and pointed toward existentialism.

That turn, Tillich told elsewhere, was furthered by two factors. One, surprisingly, was that, when still in his teens, Tillich came upon a book he regarded as the most important of all: Shakespeare's *Hamlet*, in August Wilhelm von Schlegel's translation. He learned the whole play by heart and was completely taken with its central existential question, which is expressed in German as *"Sein oder Nichtsein, das ist hier die Frage."* It may not be an accident that in the first volume of Tillich's *Systematic Theology* there is a chapter headed *"Sein und Nichtsein."*

The second factor was an event that, Tillich said in an interview, was for him a crucial life-changing experience. During the whole First World War, Tillich was a chaplain with the troops, often at the front.

One night, in a terrible battle, he saw all around him his friends and comrades die miserably of their wounds. He said about this, "My eyes were opened forever to the negative side of life. My philosophical thinking went from idealism to existentialism. . . . [I now saw] the human predicament, with its despair, guilt, anxiety, emptiness, meaninglessness, death, as seen by modern novelists and artists." "My world and idealist philosophy collapsed." His reading now included Marx and Freud, and he realized that his work was drawing on "competitive motives of thought," on both his earlier and his later sensibilities, resulting in "a certain inconsistency and indefiniteness of terminology."

During his years in Berlin, Tillich published in 1923, at age thirty-seven, his first large book, *Das System der Wissenschaften, nach Gegenständen und Methoden* (The system of the sciences according to objects and methods), with the accent strongly on the "Das." Not only was it a step to his later, second system, that of theology. We can see here already Tillich in the firm grip of a zeal for synthesis on behalf of a great mission. Intending to encompass all knowledge under three headings—*Denken, Sein,* and *Geist*—Tillich deals here essentially with the systematization of all cognitive disciplines, as he had been doing in his lectures. Therefore the book deals with logic, mathematics, phenomenology, the empirical sciences from physics to geology and the life sciences, on to psychology, sociology, history, art, law, metaphysics, ethics, philosophy, and of course theology. A main purpose of the book was to find a place for theology and to show that every field can have a theological component.

In his introduction to the book, Tillich explained that he saw it as his duty to provide "an overview of the whole of knowledge" and also to put this schema in the service of necessary social change. No wonder some of his colleagues lashed out at him.

The Fateful Meeting

As we are about to turn now to Einstein's analogous passions, we come upon a happy surprise. I said earlier that we have no sure evidence that Einstein and Tillich met personally in Berlin, but there exists at least one documentable joint appearance of those two at a meeting that had important and unexpected consequences for each of them. There exists a

Figure 1. Group of lecturers at Davos in 1928, including Einstein *(arrow at left)* and Tillich *(arrow at right)*. Image Archive ETH-Bibliothek, Zurich.

photograph (Fig. 1) that shows Einstein and Tillich at a pleasant gathering of about two dozen persons, Tillich at age forty-one, Einstein just entering his fiftieth year.

The place is Davos in Switzerland, and the time is Sunday, the eighteenth of March 1928. This is a group of well-recognized intellectuals from many fields of study. They had come from Germany, France, Austria, and Switzerland, during the spring break at their universities. It is a year before the start of the Great Depression, a time when one could still hope for a civilized century.

Davos was small then, not yet today's fashionable sports and congress center for the Masters of the Universe. The group was meeting there on a humanitarian mission. Davos was then, as it had been for decades, the site of many sanatoria for patients suffering from pulmonary tuberculosis, who hoped to be cured by the good mountain air and sunlight. The place had already gained extra attention since the publication, four years earlier, of *The Magic Mountain*, by Thomas Mann (later to become another refugee of the class of 1933). In short, we are looking at the Zauberberg, to which came Hans Kastorp and his doomed cousin, Joachim Zimmens, two typical young students whose general isolation in the sanatoria, far from the outside world of action and ideas, only worsened their condition.

This common fate of the patients was precisely why so many prominent academics had come to Davos in 1928. They were starting a month-long Alpine University. Their unselfish mission was to give lectures and hold discussions to enlighten and cheer up the young people from the sanatoria. The lectures, in German and French, were on science, philosophy, literature, jurisprudence, and sociology: six or seven of these every day, six days a week, from forty-five lecturers when all had arrived, including people like Lucien Lévy-Bruhl and Jean Piaget. The audience consisted of some 360 students and 400 others from the environs, all crowding into the Grand Hotel Curhaus.

Einstein gave the inaugural lecture, on "The Fundamental Concepts of Physics in Its Development," and he remained fully engaged during the time of his stay—perhaps too much so. He attended other professors' lectures assiduously, met with individuals and small groups, even played his violin at a chamber concert, to help raise funds for this new university. Tillich gave two lectures, one on "Religion and Culture" and the other on "The Religious Knowledge."

Little did Tillich and Einstein know that this excursion would, for each of them, be a turning point in his life. Tillich had come from his professorship at the University of Dresden. He had by now published about a dozen books and many articles, ranging from Schelling and Schleiermacher to religious socialism. This last was to be understood as socialism studied from a religious point of view. It was also intended to lead to activities improving social conditions, a goal not shared by other prominent Protestants, such as Karl Barth, who Tillich complained "virtually ignored the social situation." Einstein, on his side, was now at the height of his fame, chiefly thanks to his general relativity theory; but, like Tillich, he too was always ready to throw himself into the social problems of the day, often to the dismay and disapproval of his nearer colleagues.

At Davos, Tillich was in high demand, surrounded by students and colleagues. His former teacher, Fritz Medicus, who saw and heard him at the meeting, wrote that Tillich was clearly "the coming man in philosophy." One of Tillich's talks at Davos was scheduled to follow one by a speaker who had given the rather pessimistic prediction that civilization was declining on the exhausted soil of Europe. Tillich rose and objected. The religious person, he held, is used to finding himself in a crisis. That may even help him to avoid false certainties, and lead him to turn to the necessary and reliable certainty in God.

In the audience were two psychologists from the University of Frankfurt. They decided right then and there to propose that Tillich be called to the University of Frankfurt—and that in fact happened the next year, to Tillich's joy. Frankfurt was then a great place for a theologian, since it was the home also for the outstanding Jewish theologians Martin Buber, Franz Rosenzweig, and their brilliant student Nahum Glatzer.

In terms of intellectual production and influence, Tillich's years at Frankfurt were to be arguably his best so far. As Adams remarked about Tillich's works during those days, "We see here the fundamental impulses that pervaded Tillich's whole career. He wished to make the prophetic and sacramental, the theological and the philosophical relations relevant to the present historical situation, and he did this by means of a constant dialogue with the creative and critical figures of past and present."

In short, when Tillich left Davos and went on to his post in Frankfurt, he could consider his kindness to the Davos students well rewarded.

As for Einstein, his vigorous participation in Davos also had long-term consequences, but of a very different kind. His biographer and son-in-law Rudolf Kaiser wrote that Einstein had been persuaded to come to Davos because of his concern for those sick students, lying there without intellectual challenge. But Einstein himself did not feel physically well in the beginning of that year. As Kaiser revealed, "In Davos started [Einstein's] severe heart disease, which kept him chained to his bed for a long time." It was a debility that he was to suffer from later on.

I have said nothing yet about the inaugural lecture Einstein gave at Davos. The body of the lecture was essentially a deep bow to Schopenhauer's determinism, which Einstein had accepted from first to last.

Einstein in that year was preoccupied by the debate about the role of causality, a debate caused by the rise of the so-called Copenhagen interpretation of quantum mechanics of Bohr, Heisenberg, Born, etc. The remarkable success of their new physics was based on their thematic belief that natural phenomena at the atomic scale were not classically causal but indeterminate, probabilistic. Einstein then, and for the rest of his life, was certain that he could rely on what he called in that Davos lecture "my scientific instinct"—namely, that the new quantum mechanics was a temporary phase; that, ultimately, one would interpret "events as necessary and fully under the law of causality"; and that this law had been "divined by the great materialists of Greek antiquity." Indeed, his own theory of relativity was, he said, "nothing more than a further consequential development of the [older] field theory," based on causality.

If Einstein had not said anything else in that lecture, which Tillich surely attended, all might have been well between them. But Einstein had included another paragraph in his lecture. In fact, he had started the whole talk with a long-held opinion: that scientific triumphs throughout history, which were based on strict causality, showed the uselessness of seeking "to refer all that happens to the exercise of will on the part of invisible spirits." Moreover, he characterized that belief as worthy only of "primitive man."

Conflicting Views of the Absolute

We can only speculate what Tillich thought about Einstein's comments. Perhaps his reaction is revealed in an early version of a paragraph that appeared in his later book, *Dynamics of Faith:* "Scientific truth and the truth of faith do not belong to the same dimension of meaning. Science has no right and no power to interfere with faith, and faith has no power to interfere with science. One dimension of meaning is not able to interfere with another dimension."

Let us now take a few minutes to look more closely at Einstein, in preparation for the coming conflict with Tillich.

Einstein's urgent mental and metaphysical compunctions, his own ultimate concerns, were hinted at in this question from an article of 1916: "What goal will and can be reached by the science to which I am dedicating myself?" His answer: "To dedicate oneself to what is essential, as against what is based only on the accident of development." Under this self-imposed demand, he had turned to extending his original theory of relativity of 1905 into his general theory, or rather, as he significantly called it at first in print, to his "generalized" theory of relativity.

I have written about the great significance to Einstein of this term, *generalized*. (See Chapter 1.) Here I need only summarize by noting that in his publications, and especially in his letters from 1899 on, Einstein spoke again and again of what he confessed to be the strongest moving force in his intellectual life. "I am driven by my need to generalize," he wrote to his friend Willem de Sitter. His relativity theory of 1905—stunning though it was, and is to this day—was unsatisfactory for Einstein himself, because it did not apply to accelerated frameworks and gravitation. Working himself nearly to physical breakdown in order to generalize the special relativity theory, he produced by 1916 what has ever since been widely acknowledged an almost superhuman accomplishment, the general theory. Now Einstein was able to apply the theory to the whole cosmos. One section of his popular book of 1917 on relativity had the heading "Consideration about the World as a Whole." Nothing less. In an article of the same year, applying his theory to cosmology, he claimed that the theory allowed him to calculate the density and size of the universe itself. Upon the success of the experimental test of the theory in 1919, one could almost feel that with his theory Einstein could answer the key question, "what holds the world together in its innermost," the challenge that had obsessed Goethe's Faust. Compared with the impetus that drove Einstein, most problems of other scientists, their other concerns, seemed to sink into relative insignificance.

Einstein was confident of his success even before the 1919 test of his theory. He had declared, in a widely read speech of 1918, that the supreme task, the highest duty, of physicists is to seek the most universal elementary laws from which, by pure deduction, the whole world picture can be achieved. And he confessed there that "the longing to behold this pre-established harmony, requiring inexhaustible patience and perseverance, can only come from a 'state of feeling' akin to that of a religious worshipper or one who is in love."

On this point, Tillich would have agreed. On the final page of his book of 1923 he had called the pursuit of the science a spiritual act. He repeatedly used the trio of Eros, passion, and scientific sobriety to describe the necessary mindset for reaching toward the ultimate. Indeed, to Tillich, the pursuit of such exalted aims was what he called the "religious element in the whole intellectual enterprise," and he defined religion as "the state of being grasped by an ultimate concern." Einstein, for his part, analogously wrote that the perception in the universe of "profound reason and beauty constitute true religiosity."

After the publication and successful test of the general theory, scientists and much of the public made Einstein into a veritable icon of genius, his theory a feat of almost mystic revelation. So it was inevitable that, just as for Tillich's early work, a backlash would not be long in coming. I leave aside, in Einstein's case, the vicious anti-Semitic attacks that made him flee from Berlin in 1922 for a round-the-world trip. After Nazi gangs had assassinated his friend, Foreign Minister Walther Rathenau, Einstein's own name was found on their list, as a next one to be killed for his so-called Jewish physics and his social views.

Among those who objected to Einstein's work, though less violently, were also conservative theologians, who thought Einstein had brashly invaded their territory. To them, the proper answer to the Faustian question what holds the world together was not to be found in Einstein's bunch of tensor calculus equations, but in the presence and grace of God. Most famously, Boston's Cardinal O'Connell charged that Einstein's view of space and time is "a cloak beneath which lies the ghastly apparition of atheism." The journal *Commonweal* published numerous articles and editorials against Einstein. It may have been such attacks that decided Einstein to write his remarkable series of essays on science and religion, beginning in 1930.

Another reason for the backlash against Einstein's work was a semantic sort of time bomb, which had been ticking away ever since his theory's early form of 1905. I am speaking of the term *relativity* itself. Einstein in fact had not called his publication a relativity theory, and he did not use the term himself for years in the titles of his publications. He adopted it eventually only after most scientists, starting with Max Planck in 1906, had freely used that term to refer to Einstein's work. Einstein himself never considered his theory a revolutionary act but merely, as he put it to his friend Conrad Habicht in 1905, a point of view "making use of a modification of the theory of space and time." He saw his work as an act of simplification, of generalization, hence an aesthetically more pleasing way to think about physics.

Significantly, he called the work his "Maxwellian Program." And he complained in letters to friends that a correct term for his work would at best have been *Invariantentheorie,* a theory not of relativity at all but of the opposite—of invariance, of constancy. After all, the whole point of his theory was to find a way to rewrite the known laws of physics so as to make them for the first time independent of the relative vantage points of different observers. Moreover, a key postulate in the theory was Einstein's declaration of the absoluteness of the speed of light, regardless of the relative motion of the observer.

So it would have been not unreasonable for physicists to call Einstein's work the theory of absolutes. Of course, as he later said in resignation, it was now too late to change the terminology. The semantic time bomb exploded in the 1920s, with collateral damage to this day. The very existence of the theory of relativity has often and wrongly been held to illustrate, and in extreme cases even to be responsible for, the perceived relativization of ethics and other common values.

Einstein's long struggle with religion has been amply documented (see Chapter 1), starting with his remark on the very first page of his "Autobiographical Notes" that as a child, although "the son of entirely irreligious [Jewish] parents," he came "to a deep religiosity up to the age of twelve." One may surmise that we have here a rebellion against parental beliefs, ironically analogous to Tillich's story.

Although Einstein soon turned his back on organized religion, the seeds of his youthful religiosity flourished in his later years, partly under the influence of Spinoza's *Ethics,* one of his favorite books. His religious views are thought even to have penetrated into choices he made in his physics. Up to about 1930, this part of his personality was kept in low profile. Then, between 1930 and 1948, he published several widely discussed articles on religion and science. The first of these, called simply "Religion and Science," set forth his idea that what he called "the cosmic religious feeling" is the most advanced and only acceptable stage of religion. Elaborating in these essays on his earlier remarks, such as his brief ones at Davos, Einstein explained that the concept of a personal God was an anthropomorphic remnant of primitive times, of a "religion of fear." This primal urge had to be abandoned in favor of a Spinozistic feeling of awe and "sense of 'wonder'" at the rationality and beauty of the universe. Moreover, as one who believed in the "universal operation of the law of causation," Einstein could not entertain "for a moment," as he said, "the idea of a being who interferes in the course of events"—such as causing prayers to be answered or miracles to occur. Einstein concluded that "serious scientific workers are the only profoundly religious people."

More ideas along this line were to come from his pen shortly. A key event in our story is Einstein's essay entitled "Science and Religion." It appeared as part of a remarkable symposium on the topic "Science, Philosophy and Religion in Their Relation to the Democratic Way of Life." That symposium, held in New York in 1940, was convoked by a stellar group of intellectuals from a wide range of fields. Both Einstein and Tillich signed the call for the meeting. The main hope of this exercise was to arrive at some sort of unity among all their different fields of knowledge, in the service of preventing civilization from being undermined by unnecessary disputes, just when totalitarianism was sweeping over Europe. That symposium volume is a fascinating document.

This is where Tillich re-enters our story. When Tillich read Einstein's essay in that volume, he felt deeply troubled by it. Einstein had reemphasized his concept of cosmic religion. As he put it, a main source of the conflicts between the spheres of religion and of science lies in this concept of a personal God. Because Einstein believed that the law of causality applied to all physical events, it was inconceivable to him that a "Divine Will exists as an independent cause of natural events." He called on all "teachers of religion" to "have the stature to give up the outdated doctrine of a Personal God, that is, give up that source of fear and hope which in the past placed such vast powers in the hands of priests." He added that it was only the perception of the "grandeur of reason incarnate in existence" which "appears to be religious in the highest sense of the word." In this way, science "purifies the religious impulse of the dross of its anthropomorphism" and "contributes to a religious spiritualization of our understanding of life." Strong words.

While Tillich was of course by no means of the conservative camp, he thought that some response to Einstein's words was called for. Within two months of the publication of Einstein's essay, Tillich issued his reply, entitled (in translation) "Science and Theology: A Discussion with

Einstein" (for "discussion" Tillich used *Auseinandersetzung*, meaning also to put apart, to separate, to dissolve a partnership). Precisely because that negative critique of the notion of a personal God had been propagated by a man whom Tillich regarded as the great transformer of our physical worldview, and to whom he had often referred in admiring terms, these remarks could not be left unchallenged. In his published reply, Tillich said rather sharply that Einstein had not understood the meaning of his own words. For when Einstein had written about that awe at the "grandeur of reason incarnate in existence," Einstein had failed to notice that modern theology calls this "experience of the numinous" simply "the manifestation of the ground and abyss of being and meaning." And in attacking the conception of the personal God, Einstein was only railing against an old, out-of-date mixture of mythological and rational elements, even an "unclean" one. "No criticism of this distorted idea of God can be sharp enough." Einstein had not noticed, Tillich added, that "God" is a symbol, that the predicate "personal" can be said of the Divine "only symbolically or by analogy, or if affirmed and negated at the same time." And that symbol of God was needed for man's existence: "For as the philosopher Schelling says: 'Only a person can heal a person.' This is the reason that the symbol of the Personal God is indispensable for living religion. It is a symbol, not an object."

And there was more. Tillich's suspicion now focused also on the concept of relativity itself. Traces of that suspicion in Tillich's mind can be found years before the Davos meeting, for example in Tillich's essay of 1924, entitled "The Tension of the Absolute versus the Relative in the Philosophy of History." There, Tillich condemned relativism in history, along with Marxism and positivism. And writing four years after Davos (in "The Religious Situation," 1932), Tillich had referred directly to the "modern theory of relativity," which eliminates "every absolute point of reference." To be sure, he said there, the theory of relativity "has revealed more clearly than was previously apparent the infinity of existence." But Tillich warned that the theory "shrouds the true nature in deeper mystery than before."

Toward the end of his life, Tillich wrote a set of essays, published posthumously in 1967 with the forthright title *My Search for Absolutes*. One chapter has the heading "Absolutes in Human Knowledge and the

Idea of Truth," which draws on a basic theme in the writings of Schelling. In that chapter, Tillich confides as follows: "My choice of this subject was made out of a feeling of uneasiness—uneasiness about the victory of relativism in all realms of thought and life today . . . a total victory." "The sea of relativities . . . threatens to overwhelm us." Although Tillich was not referring directly to Einstein's work, he was bothered by what he called "the great spectacle of scientific relativism. . . . But what we have here is a game," because scientists are now dealing not with reality but with "models." The same relativism, Tillich said, can be found in contemporary positivistic philosophy and "in the growth of ethical relativism. . . . [Also], there is the great and increasing relativism in . . . religion, . . . in the secularist criticism of religion."

Against this "stream" of relativisms, Tillich wrote, he stood for Absolutes. They "make language possible, understanding possible, and truth possible." Absolutes are at the bottom of "the moral imperative." Indeed, "the experience of the Absolute-itself is experience of the holy, the sacred."

When I read this, and considered that these two men had met in Davos, and that Tillich had thought for a long time about Einstein and relativity, for better or worse, I decided to look into the Einstein Archive. That collection contains, among its 45,000 documents, Einstein's correspondence with a huge number of scholars, scientists, and other intellectuals. I searched there for any letter exchange between Tillich and Einstein. There are only three mailings, all from Tillich, and none was addressed to philosophical issues. One is an appeal to join in helping yet another refugee from Nazism, another is a request to join an organization of self-help for German émigrés, which Tillich headed for fifteen years, and the third is Tillich's analysis in 1942 of the war aims of the Allies. The contents of all these documents were fully in accord with Einstein's well-known views. But we do not know whether Einstein replied to any of them. On that, the Archive is silent.

Opportunity Lost and Found

It had started so auspiciously, with those great minds assembled on that meadow on the Zauberberg in 1928, having come together with a common sense of service. Our two protagonists had so much in common. They had been exposed to the same holistic cultural inheritance, as well as the same dangers. Each worked seriously in his own way on an integration, unification, and reconciliation of different areas of culture, including science and religion. Each had produced a commanding system in his field. Each had carefully and with much pain developed his system over a long period, making this self-appointed task the very center of his life. And when we look at their main legacies, there stands out forcefully one common theme: The quest for the unification of apparent irreconcilables. For Einstein it was, as we have seen, the quest to bring together the major fields of physics, and from there to construct for himself a religion inseparable from his cosmological thoughts.

Tillich, author of several works that attempted to find synthesis among disparate ideas, had written in the foreword of the first edition of his 1923 book, *Das System der Wissenschaften:* "All sciences [*Wissenschaften*] function in the service of one truth, and science collapses if it loses the sense of connection of the whole." One of his courses at Harvard had the breathtaking title "Religion, Art, and Science." In his search for a synthesis, Tillich was trying to develop a theological and religious position as ambitious and advanced as what Einstein was attempting in his own field. Tillich called his work once "the experiment to which my whole life is dedicated, the reunion of what eternally belongs together but what has been separated in history."

So we may be tempted, finally, to think that these two kindred spirits might well have developed a mutual elective affinity, if only they had been brought together for heart-to-heart discussions. One can imagine the conversations between these two culture-carriers of the old sort, both having inherited much from nineteenth-century sensibilities, including a sensitivity to what has been called the vocabulary of the scientific sublime, "reverence and awe, reason and progress." That a personal God is a symbol instead of a mere anthropomorphic fantasy might have been acceptable to Einstein. He might not have objected to Tillich's definition, "The ultimate concern is unconditional . . . total." In turn, it would have appealed to Tillich that Einstein's relativity was as much a search for absolutes as was Tillich's *Systematic Theology;* conversely, Einstein would of course have agreed to oppose the false uses of the word *relativity.*

But we must not dream up a "happy ending." Protagonists of such exalted stature tend to persist in their life's program with a certain stubbornness. They tend to cling fiercely to their thematic presuppositions.

In fact, that is part of their strength. Einstein and Niels Bohr, too, while deeply respecting each other, never deviated, during their long debates, from their divergent views.

The differences we find between Einstein and Tillich are for us in fact an advantage: They sharpen our understanding of the respective intellectual landscapes of the two men. And the similarities in their ambitious programs reveal to us two exemplars of true Questors who devoted their lives to complementary searches, each animated by his own ultimate concern.

References

Adams, James Luther. *Paul Tillich's Philosophy of Culture, Science, and Religion.* New York: Harper and Row, 1965.

Brauer, Jerald C., ed. *The Future of Religions / Paul Tillich.* New York: Harper and Row, 1966.

Church, F. Forrester, ed. *The Essential Tillich.* New York: Macmillan Publishing Co., 1987.

Einstein, Albert. "Fundamental Concepts of Physics and Their Most Recent Changes." *St. Louis Post-Dispatch,* December 9, 1928. [Text of Einstein's lecture at Davos.]

Einstein, Albert, et al. *Living Philosophies.* New York: Simon and Schuster, 1931.

Halter, Ernst. *Davos, Profil eines Phänomens.* Zurich: Offizin Zürich Verlags, 1994.

Kimball, Robert C., ed. *Theology of Culture / Paul Tillich.* New York: Oxford University Press, 1970.

Nemerov, Howard. *The Questor Hero: Myth as Universal Symbol in the Works of Thomas Mann.* 1940.

Science, Philosophy and Religion: A Symposium. Conference on Science, Philosophy and Religion in Their Relation to the Democratic Way of Life, Inc. New York, 1941.

Tillich, Paul. *Begegnungen.* Gesammelte Werke, vol. 12. Stuttgart: Evangelisches Verlagswerk, 1971.

——— *Biblical Religion and the Search for Ultimate Reality.* Chicago: University of Chicago Press, 1955.

——— *My Search for Absolutes.* New York: Simon and Schuster, 1967.

——— *Religiöse Reden.* Berlin, New York: Walter de Gruyter, 1987.

——— *The System of the Sciences According to Objects and Methods.* London and Toronto: Associated University Presses, 1981.

— 9 —

Henri Poincaré, Marcel Duchamp, and Innovation in Science and Art

Two more different personalities can hardly be imagined than the mathematician, physicist, and all-round polymath Henri Poincaré (1854–1912) and the artist Marcel Duchamp (1887–1968). Exquisitely attuned as he was to irony, however, Duchamp would have been delighted with this pairing, the more so, as will be shown, that some of his ideas on art can be traced back to Poincaré's publications.

Poincaré, a solid nineteenth-century figure, was a very symbol of the establishment (Fig. 2). He held five professorships simultaneously, as well as membership in both the *Académie des Sciences* and the *Académie Française*—a rare combination that made him in effect doubly immortal. Nor did it hurt his standing in society that his cousin was Raymond Poincaré, the prime minister and eventual president of the French Republic.

By temperament, Henri Poincaré was conservative; immensely productive, he published a torrent of nearly 500 papers on mathematics. Many of these announced fundamental discoveries in subjects from arithmetic to topology and probability, as well as in several frontier fields of mathematical physics, including the inception, in his book on celestial mechanics, of what is now called chaos theory.[1] And all this was done with intense focus and at great speed, with one work finished right after another. Moreover, justly regarding himself not merely as a specialist but as a "culture carrier" of the European sort, Poincaré also wrote with

---

1. For lists of Poincaré's publications, see Ernest Lebon, *Henri Poincaré* (Paris: Gauthier-Villars, 1912), and G. Mittag-Leffler, ed., *Acta Mathematica*, 38 (1921): 3–35. For the mathematical work that has led to chaos theory, see Daniel L. Goroff's introductory essay to Henri Poincaré, *New Methods of Celestial Mechanics,* 3 vols. (Woodbury, NY: American Institute of Physics, 1993).

Figure 2. Henri Poincaré. From the journal *Acta Mathematica* 38 (1921), frontispiece.

translucent rationality on the history, psychology, and philosophy of science, including his philosophy of conventionalism, in such semi-popular books as *Science and Hypothesis*.[2] For cultured persons all over the world, and especially in France, these were considered required reading.

2. Henri Poincaré, *Science and Hypothesis* (New York: Dover, 1952). Originally published as *La Science et l'hypothèse* (Paris: Flammarion, 1902).

Figure 3. Photograph of Marcel Duchamp and Eve Babitz at the Pasadena Retrospective in 1963. (Courtesy of Mme. Jacqueline Matisse Monnier. Photo © 1963 Julian Wasser.)

On the other hand, our Marcel, thirty-three years younger, would help in his much more secretive way to fashion the aesthetic sensibility of the twentieth century. The poet and critic Octavio Paz went so far as to write, in the first sentences of his book on Duchamp, "Perhaps the two painters who have had the greatest influence on our century are Pablo Picasso and Marcel Duchamp: the former by his works; the latter by a single work that is nothing less than the negation of work in the modern sense of the word."[3]

3. Octavio Paz, *Marcel Duchamp: Appearance Stripped Bare* (New York: Viking Press, 1978), p. 1.

As this quotation hints, Duchamp was in his way an inheritor of that strong, late-nineteenth-century movement, French anarchism. He recorded his admiration for a bible of anarchism, *The Ego and His Own*, by Max Stirner, first published in German in 1844 and translated many times.[4] Duchamp was co-founder of the New York Dada group and loved nothing more than to shock the establishment, calling himself, in one of his articles of 1915, an iconoclast. He was irreverent, brilliantly playful, casually misleading, an experimenter in eroticism in art—the very antithesis of bourgeois convention (Fig. 3). It could take him years to bring a project to fruition; he would write several hundreds of private notes on the way, and he even published several boxes of these notes in facsimile. But he also produced quietly a large oeuvre, while pretending, or at least pretending to pretend, that he was ever the most unpreoccupied, the most *decontracté* man in God's wide world.

And yet, despite all their differences on the surface, I shall try to indicate that the minds of these two very different icons, Poincaré and Duchamp, intersected like two planes in a space of higher geometry.

The "Monster of Mathematics"

First, let us get better acquainted with Poincaré. As one of his biographers, Jean Dieudonné, put it, Poincaré's teachers at the *lycée* in his native Nancy, where Poincaré rarely had to take a note in class, had early identified him simply as "the Monster of Mathematics." And that he remained to his death, at age fifty-eight in 1912, when the young Marcel was really just getting started. Poincaré was a mathematical genius of the order of Carl Friedrich Gauss. Before he was thirty, Poincaré became world-famous for his discovery, through his ingenious use of nonEuclidean geometry, of what he called the fuchsian functions.

Although Poincaré generally kept himself well informed of new ideas in mathematics, on this particular point his biographer allows himself the schoolmasterly remark that "Poincaré's ignorance of the mathematical literature, when he started his researches, is almost unbelievable. . . .

---

4. Quoted in Linda Dalrymple Henderson, *Duchamp in Context* (Princeton: Princeton University Press, 1998), pp. 61–62.

He certainly had never read Riemann"⁵—referring to Bernhard Riemann, the student of Gauss and the person first to define the *n*-dimensional manifold, in a lecture in June 1854. That lecture (later published) and the eventual opus of Riemann's three-volume-long series of papers are generally recognized to be monuments in the history of mathematics, inaugurating one type of non-Euclidean geometry.

For over two thousand years, Euclid's *Elements* of geometry had reigned supreme, for its demonstration that from a few axioms the properties of the most complicated figures in a plane or in three-dimensional space followed by deduction. To this day, Euclid is the bane of most pupils in their early lessons; but to certain minds his *Elements* was and is representative of the height of human accomplishment. To Albert Einstein, who received what he called his "holy" book of Euclidean geometry at age twelve, it was a veritable "wonder." He wrote: "Here were assertions ... which—though by no means evident—could nevertheless be proved with such certainty that any doubt appeared to be out of the question. This lucidity and certainty made an indescribable impression upon me."⁶

Galileo, too, was stunned by his first encounter with Euclid. It is said that as a youngster, destined to become a physician, he happened to enter a room in which Euclidean geometry was being explained. It transfixed him and set him on his path toward finding the mathematical underpinnings of natural phenomena. To Immanuel Kant, of course, Euclidean geometry was such an obvious necessity for thinking about mathematics and nature that he proposed it as an exemplar of the *synthetic a priori* that constituted the supporting girder of his philosophy.

By the early part of the nineteenth century, however, a long-simmering rebellion came to a boil against the hegemony of Euclidean geometry, and especially its so-called fifth axiom; that one implies, as our schoolbooks clumsily state it, that only one line can be drawn through a point next to a straight line which is parallel to it, both of them intersecting only at infinity. Here, the main rebels were Riemann, the Hungarian János Bolyai, and the Russian Nicolai Ivanovich Lobachevsky,

---

5. Quoted in the *Dictionary of Scientific Biography*, vol. 11 (New York: Charles Scribner's Sons, 1973), p. 53.

6. Einstein, quoted in Paul A. Schilpp, ed., *Albert Einstein: Philosopher-Scientist* (Evanston, IL: Library of Living Philosophers, 1949), p. 9.

who all produced geometries in which the fifth axiom was disobeyed. As to Poincaré's initial ignorance of that particular literature, ignorance can be bliss, not merely because reading Riemann's turgid prose is no fun, but chiefly because Poincaré's approach was original with him.

Discovery and the Unconscious

It is easy to document Poincaré's intellectual conservatism. For example, he refused to accept Einstein's relativity theory, and he clung to the notion of the ether. But we have learned from other examples that those innovators, both in science and the arts, who are far above our own competencies and sensibilities are not easily pinned down intellectually. They tend to have what seem to us to be contradictory elements, which nonetheless somehow nourish their creativity. Thus there is also plenty of evidence of Poincaré's willingness to face the force of sudden change. That trait emerged strikingly in his ideas about the psychology of invention and discovery. The reference here may be familiar, but it is so startling that it deserves nevertheless to be mentioned—and read with the eyes of the artists of the time.

I refer to the famous lecture on "L'invention mathématique" which Poincaré gave in 1908 at the *Société de Psychologie* in Paris. The fine mathematician Jacques Hadamard remarked on that lecture that it "throw[s] a resplendent light on relations between the conscious and the unconscious, between the logical and the fortuitous, which lie at the base of the problem [of invention in the mathematical field]."[7] In fact, Poincaré was telling the story of his first great discovery, the theory of fuchsian functions and fuchsian groups. Poincaré had attacked the subject for two weeks with a strategy (typical in mathematics) of trying to show that there could not be any such functions. Poincaré reported in his lecture, "One evening, contrary to my custom, I drank black coffee and could not sleep. Ideas rose in crowds; I felt them collide until pairs interlocked, so to speak, making a stable combination."[8] During that sleepless night he found that he could in fact build up one class of those

---

7. Jacques Hadamard, *The Psychology of Invention in the Mathematical Field* (New York: Dover Publications, 1945), p. 12.

8. Ibid., p. 14.

functions, though he did not yet know how to express them in suitable mathematical form.

Poincaré explained in more detail: "Just at this time, I left Caen, where I was living, to go to a geological excursion. . . . The incidence of the travel made me forget my mathematical work. Having reached Coutance, we entered an omnibus to go someplace or other. At the moment when I put my foot on the step, the idea came to me, without anything in my former thoughts seeming to have paved the way for it, that the transformations I had used to define the fuchsian functions were identical with those of non-Euclidean geometry. I did not verify the idea; I should not have had time, as, upon taking my seat in the omnibus, I went on with a conversation already commenced, but I felt a perfect certainty. On my return to Caen, for conscience's sake, I verified the results at my leisure."[9] Poincaré analyzed such intuitions in these terms: "Most striking at first is this appearance of sudden illumination, a manifest sign of long, unconscious prior work. The role of this unconscious work in mathematical invention appears to me incontestable." "It seems, in such cases, that one is present at one's own unconscious work, made particularly perceptible to the overexcited consciousness."[10]

Hadamard collected a number of similar reports of "unconscious" ideas, where out of continuous incubation in the subconscious there appears—in a discontinuous way, in a rupture of startling intensity—a conscious solution. He mentioned Gauss himself, who spoke of such a rupture as "a sudden flash of lightning," and similar observations by Hermann V. Helmholtz, Wilhelm Ostwald, and Paul Langevin. Nor should we forget Mozart, who spoke about the source of his musical thoughts as follows: "Whence and how do they come? I do not know, and I have nothing to do with it." We also saw this process at work for Enrico Fermi (Chapter 5).

Poincaré himself confessed puzzlement about the source of his ideas. In the full text of Poincaré's talk of 1908, soon widely read in chapter 3 of his popular book *Science and Method* (1908), he confessed, "I am absolutely incapable even of adding without mistakes, [and] in the same way would be but a poor chess player."[11] But he went on to report hav-

9. Ibid., p. 13.
10. Ibid., pp. 14–15.
11. Henri Poincaré, *Science and Method* (New York: Dover, 1952), p. 49. Originally published as *Science et méthode* (Paris: Flammarion, 1908).

ing "the feeling, so to speak, the intuition of this order [in which the elements of reasoning are to be placed], so that I can perceive the whole of the argument at a glance. . . . We can understand that this feeling, this intuition of mathematical order, enables us to guess hidden harmonies and relations. . . ."[12] To be sure, after intuition comes labor: Invention is discernment, choice." But for that, priority must be granted to "aesthetic sensibility in privileging unconscious phenomena, beauty and elegance."[13] How congenial this must have sounded to the artists among his readers!

Poincaré had earlier discussed the nature of discovery, especially in his book *Science and Hypothesis*. A point that particularly struck home at the time was that concepts and hypotheses are not given to us uniquely by nature herself but are to a large degree conventions chosen by the specific investigator for reason of convenience and guided by what he called "predilection"[14] (what I would call themata). As to the principles of geometry, he announced his belief that they are only "conventions."[15] But these conventions are not arbitrary; eventually they must result in a mathematics that "sufficiently agrees" with what we "can compare and measure by means of our senses."

Part II of *Science and Hypothesis*, consisting of three chapters, was devoted to non-Euclidean and multi-dimensional geometries. In those pages there are neither equations nor illustrations but instead great attempts at clarification by analogy. To give only one widely noted example: In attempting to make the complex space of higher geometry plausible, Poincaré introduced a difference between geometric space and conceptual or "representative" space. The latter has three manifestations: visual space, tactile space, and motor space. The last of these is the space in which we carry on our movements, an idea that led him to write, in italics, "*Motor space would have as many dimensions as we have muscles.*"[16]

---

12. Ibid., pp. 49–50.
13. Ibid., p. 396.
14. Poincaré, *Science and Hypothesis*, p. 167.
15. Ibid., p. 50.
16. Ibid., p. 55.

The Appeal of *n*-Dimensionality

To tell the truth, these and Poincaré's other attempts to popularize the higher geometries must have left the lay reader excited but not really adequately informed. Happily, at just about that time help came by way of popularizations of these ideas by others, which further enhanced the glimmer of higher mathematics in the imagination of the young artists in Paris. One such book plays an important role in our story. In 1903, a year after *Science and Hypothesis,* there was published in Paris a volume entitled *Elementary Treatise on the Geometry of Four Dimensions: An Introduction to the Geometry of* n-*Dimensions.*[17] In it, Poincaré's ideas and publications are specifically and repeatedly invoked, from the second page on. The author is now almost forgotten, the mathematician E. Jouffret (the initial hiding his wonderful first names, Esprit Pascal). In 1906 he added a more stringent treatment, in his volume *Mélanges de géométrie à quatre dimensions.*[18] There is also a good deal of evidence that a friend of the circle of artists in Paris, an insurance actuary named Maurice Princet, was well informed about the new mathematics and acted as an intermediary between the painters and such books as Jouffret's. There is a wealth of scholarship by art historians on this case; in my opinion the best presenter of the impact, especially of Jouffret's books, is Linda Dalrymple Henderson, who has discussed it in two volumes, *The Fourth Dimension and Non-Euclidean Geometry in Modern Art* and *Duchamp in Context.*[19] I shall draw on these sources without shame.

Of course, non-Euclidean geometry had been around for many decades, Lobachevsky writing in 1829 and Bolyai in 1832. But these authors were little read even by mathematicians until the 1860s. Then, for two decades to either side of 1900, a growing flood of literature, professional

---

17. E. Jouffret, *Traité élémentaire de géométrie à quatre dimensions et introduction à la géométrie à n dimensions* (Paris: Gauthier-Villars, 1903).

18. E. Jouffret, *Mélanges de géométrie à quatre dimensions* (Paris: Gauthier-Villars, 1906).

19. See Henderson, *Duchamp in Context,* cited above, and Linda Dalrymple Henderson, *The Fourth Dimension and Non-Euclidean Geometry in Modern Art* (Princeton: Princeton University Press, 1983). See also Henderson's article, "The *Large Glass* Seen Anew," *Leonardo,* 32 (1999): 13–126.

and popular, fanned the enthusiasm about that unseen fourth spatial dimension.

This cultural phenomenon has various possible explanations. First, one must not forget that this branch of mathematics was still at the forefront of lively debate among mathematicians themselves. There were about fifty significant professionals in Europe and in the United States working in that area. Second, to laypersons, the new geometry could be a liberating concept. It hinted at an imaginative sphere of thought not necessarily connected to the materialistic physical world that had been presented by nineteenth-century science—which in any case was itself in upheaval thanks to a stunning series of new discoveries. Above all, the new geometries lent themselves to wonderful and even mystical excesses of the imagination, not least by literary and figurative artists and musicians. They included Dostoyevsky; H. G. Wells; the science fiction writer Gaston de Pawlowski; dramatist Alfred Jarry, who mentioned pataphysics in one of his plays; Marcel Proust; the poet Paul Valéry; Gertrude Stein; composers Edgard Varèse and George Antheil; the influential Cubists Albert Gleizes and Jean Metzinger, and so on—not to speak of P. D. Ouspensky and the Theosophists. Some painters were explicit about their interest. Kazimir Malevich gave the subtitle "Color masses in the Fourth Dimension" to a work in 1915, and as late as 1947 the surrealist Max Ernst produced a painting he called *L'homme intrigué par le vol d'une mouche non-euclidienne*.

The Importance of Popularizers

For French artists getting interested in these matters in the first decade or so of the new century, Poincaré's writing whetted their appetite. They also found his writings especially sympathetic, not only because of his literary and rhetorical skill but because he defended the new geometries as conveniences or conventions rather than as *synthetic a priori* or as facts of experience, and because of his opposition to reducing mathematics to logic, emphasizing instead the importance of intuition, as we have seen.

To be sure, Poincaré warned against attempts at easy visualization, saying "a person who devotes his whole existence to it may perhaps attain to a realization of the fourth dimension." But one could approach it by analogy, just as a creature in Plato's cave seeing on its wall only the shadows cast by persons moving and turning outside may induce not only their existence but also their three-dimensionality. Or one could imagine helping a two-dimensional creature, which is living entirely on a plane, begin to understand what a three-dimensional cone is like: Take a cone (say a carrot), make a large number of infinitely thin slices, cuts made through the cone this way and that, and lay the slices of various dimensions on the plane for that creature to inspect by crawling around them. It will eventually get the idea. Analogously, Poincaré described in 1902 how we might observe in three or two dimensions what a figure is like that exists only in the fourth dimension: "We can take of the same figure several perspectives from different points of view" as in rotating it, cutting it, and so on. Duchamp himself, in one of his notes on his work *A l'infinitif*, wrote: "The set of these 3-dimensional perceptions of the 4-dimensional figures would be the foundation for the reconstruction of the 4-dimensional figure."

In Duchamp's working papers, we even have evidence—so to speak, a smoking gun—that he was intrigued by Jouffret's *Treatise* on four-dimensional geometry. In his extensive notes of 1912–1914, made in preparation for constructing his ever-unfinished *Large Glass*, he wrote (Fig. 4) about a passage in Jouffret's book on how a two-dimensional shadow is cast by a three-dimensional figure. Duchamp pointed to an analogy: "The *shadow* cast by a 4-dimensional figure in our space is a *three-dimensional shadow* (see Jouffret's *Géométrie a 4 dim*, . . . p. 186, last three lines). . . . Construct all the 3-dim'l states of the 4-dim'l figure, the same way one determines all the planes or sides of a 3-dim'l figure—in other words: A 4-dim'l figure is perceived (?) through an ∞ of 3-dim'l sides, which are the sections of this 4-dim'l figure by the infinite numb. of spaces (in 3 dim.) which envelope this figure.—In other words: one can *move* around the 4-dim'l figure according to the 4 directions of continuum."[20]

---

20. Among many other quotations by Duchamp on his interest in four-dimensional geometry is a publication by Stephen Jay Gould and Rhonda Roland Shearer, "Boats and Deckchairs," *Natural History* (December 1999). A good, nonmathematical account of the early use of the fourth dimension by scientists is Alfred M. Bork, "The Fourth Dimension in Nineteenth-Century Physics," *Isis*, 55 (1964): 326–338.

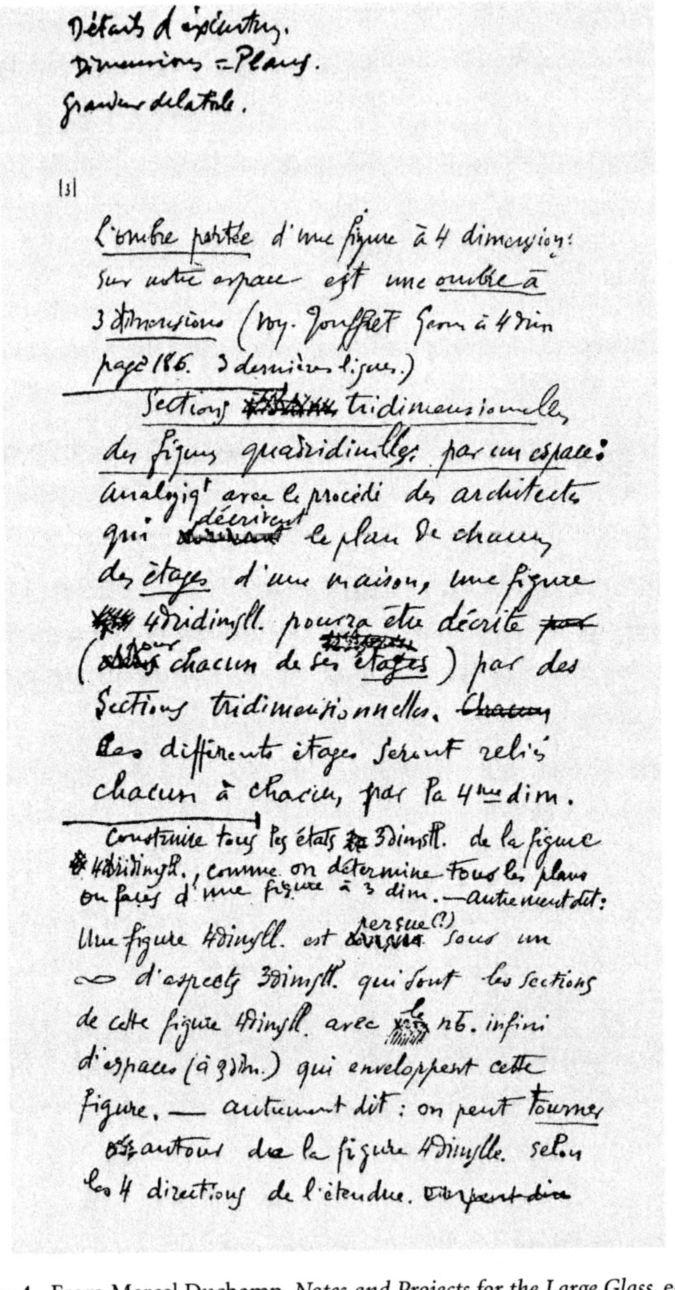

Détails d'exécution.
Dimensions = Plans.
Grandeur de la toile.

[3]

L'ombre portée d'une figure à 4 dimensions sur notre espace est une ombre à 3 dimensions (voy. Jouffret Géom. à 4 dim. page 186. 3 dernières lignes.)

Sections tridimensionnelles des figures quadridimlles. par un espace: Analogiqt avec le procédé des architectes qui décrivent le plan de chacun des étages d'une maison, une figure 4 dimlle. pourra être décrite (pour chacun de ses étages) par des sections tridimensionnelles. Les différents étages seront reliés chacun à chacun, par la 4me dim.

Construire tous les états 3dimsll. de la figure 4dimsll., comme on détermine tous les plans ou faces d'une figure à 3 dim. — autrement dit: Une figure 4dimsll. est perçue (?) sous un ∞ d'aspects 3dimsll. qui sont les sections de cette figure 4dimsll. avec le nb. infini d'espaces (à 3 dim.) qui enveloppent cette figure. — autrement dit: on peut tourner autour de la figure 4dimslle. selon les 4 directions de l'étendue.

Figure 4. From Marcel Duchamp, *Notes and Projects for the Large Glass,* edited by Arturo Schwartz (New York: H. N. Abrams, 1969. © Artists Rights Society [ARS] New York/ADAGP, Paris.)

I do not want to be misunderstood here to imply that Duchamp was simply a student and user of higher mathematics, like a pupil in a class. Rather, we must remind ourselves that artists do not need to cogitate with the same rigor as good scientists are supposed to do. Their uses of science, or of scientific fascinations like the fourth dimension, are likely to be scientifically casual, perhaps not scientific at all. And thank goodness for that! They can turn the reports of science into something new through their sensibilities. For example, some of the most adventurous pictorial artists of that period were especially impressed by the new idea that surfaces and spaces may exist with varying curvatures, which distorted the figures that moved in them, contrary to the linear perspective system that had dominated painting for centuries.

For example, consider the morphing of a body in n-dimensional space, a space which itself can be curved, as the body is transported in that space from one region to another. To prepare yourself by an analogy, look first at a part of a globe representing our Earth—say, at a state such as Kansas, which is by and large a rectangular territory, defined by two parallels of longitude and, perpendicular to them, two parallels of latitude. If we now moved Kansas up along the two longitudes, it would become more and more squeezed, ending up as a triangular slice, as the top of the state reaches the North Pole. In that spirit, one can then imagine the dramatic distortions of three-dimensional objects—the surrealistic flexibility and fluidity they would have to exhibit—if we could move them around in non-Euclidean space.

Thoughts of this kind seem to some art critics to be behind one of Duchamp's remarks concerning his masterpiece, *The Bride Stripped Bare by Her Bachelors, Even,* also known as the *Large Glass* (Fig. 5). Octavio Paz calls it "one of the most hermetic works of the century."[21] In its upper panel, a vaguely cloud-like structure represents part of the Bride. Duchamp told Richard and George Hamilton in 1959 that the Bride is "half-robot, half-four dimensional," and he invited Pierre Cabanne and André Breton to look at that part "as if it were the pro-

---

21. Paz, *Marcel Duchamp,* p. 29.

Figure 5. Marcel Duchamp, *The Bride Stripped Bare by Her Bachelors, Even (The Large Glass)*, 1915–1923. (Philadelphia Museum of Art; Bequest of Katherine S. Dreier. © 2001 Artists Rights Society [ARS], New York/ADAGP, Paris.)

jection of a 4-dimensional object."[22] There is additional evidence in the literature that Duchamp may have been serious when making these comments.

I mentioned that, according to Duchamp's notes, an aid to his imagination came from Jouffret's 1903 *Treatise*. Jouffret labored to describe indirect ways to visualize the higher dimensions that could well have been of great appeal to some artists living through that revolutionary period. For example, see Fig. 6, taken from that book. Henderson says of those illustrations that they are the "sixteen fundamental octahedrons" of an icosatetrahedroid (made up of twenty-four octahedrons) "projected on a two dimensional page, frequently being turned," to show aspects of their third and fourth dimensionalities. The second of the two Jouffret illustrations is an example of a projected "see-through view" of the four-dimensional body.[23]

At this point one cannot resist showing also some of Picasso's works of 1900–1912, such as his 1910 *Portrait of Wilhelm Uhde* (Fig. 7). Its complex construction is based, in Henderson's words, on a "variety of planes and angles seen from different points of view."[24] These have a suggestive similarity to the triangular facets in Jouffret's "see-through" figures, resulting from the projection on two dimensions of four-dimensional objects. As in other paintings made by Picasso during those years and by others in that circle, there is here an ambiguity of form, certainly a turning away from the perspective that had reigned over most painting since the Renaissance. But Henderson cautions: "In no way is a causal relationship being suggested between n-dimensional geometry and the development of the art of Picasso and Braque."[25] Among many factors that may have influenced those artists—some palpable, others almost subliminal—she notes the proof that the newly discovered x-rays furnished of the limited perceptual powers of the unaided human eye.

22. Henderson, *Fourth Dimension*, p. 57.
23. Ibid., p. 57.
24. Ibid., p. 58.
25. Ibid., p. 58.

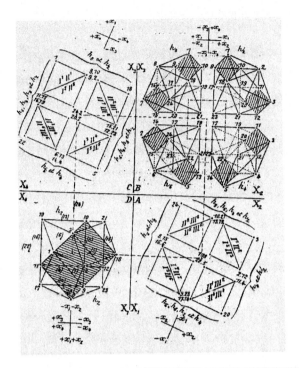
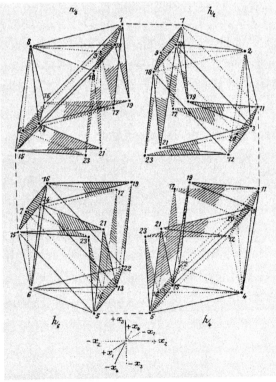

Figure 6. E. Jouffret's guide for presenting four dimensions on paper. From E. Jouffret, *Traité élémentaire de géométrie à quatre dimensions et introduction à la géométrie à n dimensions* (Paris: Gauthier-Villars, 1903), pp. 152, 153.

Figure 7. Pablo Picasso, *Portrait of Wilhelm Uhde*. (© 2001 Estate of Pablo Picasso/Artists Rights Society [ARS], New York. Photo: Giraudon/Art Resource, New York.)

Zeitgeist or Common Knowledge?

Here, however, I find myself in the middle of a mine field where huge battles are being fought by art historians. I withdraw cautiously, and look in other directions for explanations for the appearance of analogous ideas in different fields. One is tempted to seek refuge in a suggestive idea expressed by the philosopher José Ortega y Gasset. Writing in *The Modern Theme* (published in English in 1933) of the reason relativity arose when it did, for example, Ortega said the most relevant issue is not that the triumph of *one* particular theory "will influence the spirit of mankind by imposing on it the adoption of a definite route," in which innovators in many diverse fields find themselves doing analogous things. "What is really interesting," he continues, "is the inverse proposition; that the spirit of man sets out, of its own accord, upon a definite route," and that this process allows a theory to be born and to manifest itself in different guises, producing "profound variations in the mind of humanity."[26]

There, the philosopher comes perilously close to invoking the *Zeitgeist*, that dark abyss which has swallowed up all who have attempted to peer down into it. E. H. Gombrich warned of the danger in these words: "[O]bviously there is something in the Hegelian intuition that nothing in life is ever isolated, that any event and any creation of a period is connected by a thousand threads with the culture in which it is embedded. [Yet] it is one thing to see the interconnectedness of things, [and] another to postulate that all aspects of a culture can be traced back to one key cause of which they are the manifestations. [This latter viewpoint] demands that everything must be treated not only as connected with everything else, but as a symptom of something else.... But I see no reason why the study of those connections should lead us back to the Hegelian postulates of the *Zeitgeist* and *Volkgeist*."[27]

Yet, there remains in some of us the haunting feeling that something like a spirit of the times may exist. The psychologist Edwin G. Boring once wrote a charming essay on this topic in which he, a good positivist,

---

26. José Ortega y Gasset, *The Modern Theme* (New York: W. W. Norton, 1933), pp. 135–136.

27. E. H. Gombrich, *In Search of Cultural History* (Oxford: Clarendon Press, 1969), pp. 30–31. I thank one of my anonymous referees for guiding me to this reference.

tried to demystify the concept of the "*Zeitgeist* as a vague supersoul pervading and controlling the immortal body of society." He put in its place the contrary definition, that the *Zeitgeist* is simply "the total body of knowledge and opinion available at any time to a person living within a given culture."[28]

That seems to me to bend too far in the opposite direction. Just think of how the turbulent new world of ideas and means would have been able to interact with the alert soul of an artist, especially in Paris during the optimistic two decades just prior to the catastrophe of 1914. There was the miracle of electrification of the cities; the futurism screaming from new inventions such as cinema, wireless telegraphy, radio, airplanes, and automobiles. Add to it the outrageous, unforgettable, disjointed, and asymmetrical dance movements introduced by the *Ballets Russes* from 1909 in Paris, under Diaghilev; the scandalous but irresistibly dissonant explosion of Stravinsky's *Rite of Spring*, just in 1913; and not least, the stream of hot news from the laboratories, each story more spectacular than the last: x-rays, the electron, radioactivity, relativity, the nucleus, the definitive verification by Jean Perrin of the atomic hypothesis, Niels Bohr's 1913 explanation of the structure of the atom. Consider the reaction of the artist Wassily Kandinsky in his memoir about the early years of the twentieth century.[29] He had earlier experienced a block in his artistic work, but when he heard of some of those astounding novelties in science, his block vanished, he perceived "a collapse of the whole world," and so a new beginning was possible for him.

If a "spirit of the times," one expressing itself in the interaction of different parts of a culture's processes of discovery and invention, is more than a nostalgic phrase, one may perhaps hope to find a living example of it just in those last Banquet Years in Paris. Whatever seeds were planted in the artists' imaginations owing to the mathematicians' popularization of the new geometry would find good nourishment also from those other inseminating influences of that era.

---

28. 12. Edwin G. Boring, ed., *The Validation of Scientific Theories* (Boston: Beacon Press, 1956), p. 215.

29. Wassily Kandinsky, *Rückblick* (Baden-Baden: Woldemar Klein Verlag, 1955), p. 30. See G. Holton, *Einstein, History, and Other Passions* (Cambridge, MA: Harvard University Press, 2000), chap. 6, "Einstein's Influence on the Culture of Our Time."

The Alchemy of Culture

To end, a remark regarding my personal motivation in presenting this essay. Having considered the thoughts of mathematicians as they acted on artists at whatever remove, I should note that this of course is only one of many documentable examples of the interdigitation of one part of culture with another part. For example, we all know of the effect Newton's physics had on eighteenth- and early-nineteenth-century poets. What interests me even more, though I could not show it in the case presented here, is that at its best this process is a mutual one. In the very nature of science-done-at-the-highest-level, one often encounters a humanistic, philosophical challenge at the core of the scientific one. The interpenetration reveals itself through the use by scientists of metaphors and of the thematic imagination, in many cases enhanced by literature and philosophy. Niels Bohr explained that he came to his complementarity principle in part through reading Søren Kierkegaard and William James. Einstein's program of unification of physics was encouraged by his reading of Goethe and other authors of the Romantic period in literature. Similar examples can be traced in other sciences, such as the debt of Kepler's astronomy to the music theory of the day and Kurt Gödel's inspiration received from his study of neo-Platonism, Leibnitz, and Immanuel Kant.

That debt is repaid many times over, by the effect that new scientific ideas can have on very different fields. Newton's *Principia* was discussed by the Founding Fathers as containing a model for the Constitution of the United States. Similarly, after the experimental proof of the General Relativity Theory in 1919, there ensued decades of transformation and resonances of the theory among philosophers such as Henri Bergson, A. N. Whitehead, and Karl Popper, as well as among literary figures such as William Carlos Williams, Archibald MacLeish, E. E. Cummings, Ezra Pound, Thomas Mann, Hermann Broch, and William Faulkner.

Indeed, a culture is kept alive by the interaction of all its parts. Its development is an alchemical process, in which the culture's varied ingredients combine to form new jewels. On this point, I imagine that both Poincaré and Duchamp would be in agreement.

10

Perspectives on the Thematic Analysis of Scientific Thought

It is now over four decades since I first published on the concept of a thematic analysis of scientific thought (in *Eranos-Jahrbuch* of 1962). The approach laid out there and in the book published by Harvard University Press in 1973 (revised edition, 1988) has been at the heart of much of my subsequent research in the history of science, as well as having been taken up by others. It may therefore be appropriate to look back and present a personal perspective on this subject.

Thematic Origins

Given that aim, the first thing must be a confession. Studying the thematic origins and elements in scientific works—with its focus on the personal presuppositions, especially in the nascent stages—was not an activity for which I was prepared by my early education and beliefs. On the contrary. My early intellectual formation might be said to have taken place under the auspices of a kind of positivism which took for granted, in Hans Reichenbach's famous phrase, that one should be "interested not in the context of discovery, but in the context of justification."

My doctoral thesis, on experimental high pressure physics, was done under the supervision of P. W. Bridgman, who, besides being a Nobel Prize physicist, was also deeply interested and active in philosophy (see Chapter 6). His contribution came to be known under the name of operationalism, a version of pragmatism/instrumentalism, which has deep roots in American philosophy of science. His *Logic of Modern Physics* (1927) and his many subsequent articles were extraordinarily influential, especially in the United States and far beyond the natural sciences.

I also was taking courses on philosophy, including one on the philosophy of science given by the physicist-philosopher Philipp Frank. I was then asked by him to be his assistant, and we became close friends and colleagues. Frank had been one of the pillars of the Vienna Circle, and from 1938 on he brought the Unity of Science movement and logical empiricism to the United States, making Cambridge its most active center. Monthly meetings of his group took place for about a decade, attracting scholars from other parts of the United States to Cambridge. I took part in its sessions, having been invited to be the recording secretary of the group—an appropriate position for the person who was by far the youngest in the room. In addition to all these possible influences upon me, there was also the magisterial Richard von Mises in the Applied Science Department, next to ours. A powerful missionary, he had targeted me to translate into English his book on positivism. (I mustered the courage to decline that, and I am sure he never forgave me.) W. V. Quine, one of the participants, described the meetings in his autobiography as the "Vienna Circle in Exile."[1]

One might have expected that I, living in this ambiance for many years, would become a disciple of later-style positivism. What was it that prevented such a development, so contrary to the ideas of my distinguished early sponsors, colleagues, and friends? In retrospect, I see that a seed of my later interests appeared, perhaps to my own surprise, in my earliest historical study, on Johannes Kepler's torturous way to his three great laws.[2] The data Kepler had to work with—the phenomenic dimension—and the logical/analytical system he brought to bear on the inter-

---

1. I have described the genesis and fortunes of the group and its meetings in chap. 1 of my book *Science and Anti-Science* (see the reference list at the end of this chapter), and in the essay "On the Vienna Circle in Exile: An Eyewitness Report" in W. De Pauli-Schimanovich et al. (eds.), *The Foundational Debate* (Dordrecht: Kluwer Academic Publishers, 1995, pp. 269–292). Quote from W. V. Quine, *The Time of My Life: An Autobiography* (Cambridge, MA: MIT Press, 1985), p. 219.

2. It was published in the *American Journal of Physics* in 1956 under the title "Johannes Kepler's Universe: Its Physics and Metaphysics" (and reprinted in Chapter 2 of *Thematic Origins*).

pretation formed, so to speak, the *x-y* axes of the plane of discourse of his science, as we see it also in its modern form, in the published scientific papers. In addition, I found that Kepler was highly motivated and helped, or sometimes hindered, by a third component of his approach, a third *(z)* dimension, orthogonal to the other two. In his case, that third dimension was populated with hardy presuppositions that were neither confirmable nor falsifiable, not arising from the data or the theory but imposed on them by him from the outset, accepted explicitly or implicitly, until the end or until he found that they had to be replaced after all by other presuppositions.

Among these presuppositions forming the thematic framework within which he labored over his data and calculations, one finds his belief that three-dimensional Platonic bodies are a key to the explanation of distance between planetary orbits; that the old *hantisement* (obsession) of the circle was another such key; then from 1605 on, as he put it, "that the celestial machine is to be likened not to a divine organism but rather to a clockwork." And, finally, his presuppositions included the brilliant superposition of the cosmological model in terms of three themata: the universe as physical machine, the universe as mathematical harmony, and the universe as central theological order.

My growing interest in the history and philosophy of science in those years was being furthered by teaching physics courses in the new General Education program at Harvard which, not least through the interest and participation of the president of the university himself, James Bryant Conant, were based on the science history of the topics being taught.

But the most powerful force to change my intellectual career came almost by accident, not long after Albert Einstein's death in 1955. Philipp Frank had been a close friend and biographer of Einstein, his successor at the German University in Prague in 1912, at Einstein's recommendation. He suggested I might prepare some historical account of aspects of Einstein's work to be used at a memorial service. To my astonishment, I found that there was then hardly any work being done by historians of science on Einstein's fundamental advances, although there existed more than enough biographies. In Chapter 2, I recounted my introduction to the Einstein estate, my initial visit to the Institute for Advanced

Study at Princeton to see if there was some first-hand material on which to base an historical essay, and my subsequent immersion in Einstein's correspondence and manuscripts held there.

The main point of interest for my account here is that through reading Einstein's drafts of his work and copies of his correspondence, I came to understand that, in addition to his attention to data and logical/analytical tools (which Einstein himself referred to as the "empirical" and the "rational" aspects), a mainspring of his work was often a set of fiercely held presuppositions, what I had identified as thematic elements. Among the themata guiding Einstein's theory construction were clearly the following: the primacy of formal rather than materialistic explanation; unity or unification; logical parsimony and necessity; symmetry; simplicity; causality in the classical sense; completeness in the subordination of every phenomenon under the respective theory; the continuum; and of course constancy and invariance.

Once seen clearly in the nascent stages, these thematic components can sometimes, with some effort, be observed also in the published papers, although Einstein, like most scientists, took care to keep such motivating, seemingly metaphysical aspects of his work out of view. The hold of these on his wide-ranging imagination explains why he would obstinately continue in a given direction even when testing against experience (as in both special and general relativity theory) was initially unavailable or difficult, and why, on some occasions, he held fast even contrary to recent experimental results. It explains also what to some commentators have been persistent puzzles: why Einstein was hostile to theories well supported by correlation with experiments but based on thematic presuppositions opposite to his own (as in the case of the quantum mechanics of Niels Bohr's school, with its emphasis on discontinuity, probabilism, and incompleteness in the description of phenomena; and why to him Michelson's experiments were not crucial in the genesis of the relativity theory).

Philosophers of science had overlooked Einstein's pilgrimage from a self-confessed follower of Ernst Mach's ideas to a position much closer to that of Max Planck. Indeed, one of these, a former student of Hans Reichenbach, chided Einstein in print in no uncertain terms for not having acknowledged what this critic took to be a crucially necessary experimental support for Einstein's initial proposal of the principles of relativity theory.

My first public attempt to present my view of Einstein's work came in 1959 at the International History of Science meeting in Barcelona.³ A more extensive investigation of the powers and limits of thematic analysis was delivered in 1961 at one of the Eranos conferences in Ascona, on Lago Maggiore. That seemed to me the safe place to try out my ideas at length, because those annual conferences had two charming properties: In the morning one could lecture for as much as three hours before a small but excellent audience (typically including Gershom Scholem, Herbert Reid, Karl Deutsch, Hans Richter, and other scholars and artists) who then participated, in the afternoon, in another lengthy period of questions and discussion. Secondly, the proceedings were published in a volume of very small circulation—which accounts for my daring to entitle my talk "About the hypotheses on which are based the natural sciences,"⁴ boldly referring to the famous *Habilitationsvortrag* of Bernhard Riemann, who had died a century earlier in a town on the same lake.

You might wonder, as indeed I did, what my colleagues back in Cambridge would think about my conceptions, which I shared with them in manuscript, as a sort of declaration of independence; but they were generous and forgiving. In any case, I felt that I was now launched on fruitful work, publishing examples of thematic analysis in case studies that were collected in 1973 as *Thematic Origins of Scientific Thought*, which focused on aspects of the contributions by Kepler, Newton, Bohr, and Einstein. In later publications, I also turned to others, including Galileo, Fermi, R. A. Millikan, and Heisenberg.⁵

---

3. "Continuity and Originality in Einstein's Special Relativity Theory," *Actes du IXe Congrès Internationale d'Histoire des Sciences* (Barcelona, 1959).

4. "Über die Hypothesen, welche der Naturwissenschaft zu Grunde liegen," in Adolf Portmann (ed.), *Eranos-Jahrbuch XXXI/1962* (Zurich: Rhein-Verlag, 1963, pp. 351–425).

5. The concept of subjecting scientific work to thematic analysis was discussed and taken up by others, as mentioned in the Postscript to the 1988 edition of *Thematic Origins* (such as R. K. Merton, Erik Erikson, Roman Jakobson, Robert Nisbet, Helge Kragh, John Losee), and more since that date, including Peter Galison and S. S. Schweber.

The Thematic Approach

Turning now from an autobiographical sketch to a review of the main points of the work itself, it may be helpful, especially for those who have not followed the books and articles in which the role of themata were dealt with, to outline the thematic approach, not least in order to clarify the distinctions with entirely different conceptions, such as archetypes, paradigms, or Kantian categories.

To summarize essentials and to set the stage, consider that when scientists publish the results of their work, they are submitting them for acceptance into what could be called *public science*. They generally make quite sure to cast a veil over the prior stage of their effort, the scientist's individual activity during the nascent period, which deserves the term *private science*. The common error of using the word *science* without making this distinction can show up glaringly when the historian of science tries to understand the motivation of scientists for pursuing their research problems, the original choice of their conceptual tools, or the treatment of their data. In all these cases, one may discover that during the nascent, "private" period of work, some scientists, consciously or not, use highly motivating, very general thematic presuppositions. But when the work is then proposed for entry into the "public" phase of science, these motivating aids tend to be suppressed, and even disappear from view. Even though thematic notions arise from a deep conviction about nature, on which the initial proposal and eventual reception or rejection of one's best work may be based, they are not explicitly taught, and they are not listed in the research journals or textbook. That has certain advantages, insofar as silence about personal motivations and thematic preferences avoids any deep, unresolvable disputes in the public phase. Consensus is more easily reached if thematic elements are kept out of sight.

Some modern philosophers of science, particularly those tracing their roots to empiricism or positivism, go further and assign "meaning" to any scientific statement only insofar as the statement can be shown to have phenomenic and/or analytic components in this plane. To be sure, this policy has freed science from innate properties, occult principles, and other tantalizing, "metaphysical" notions that cannot be resolved into components along those x and y axes. It was in part for the sake of this advantage that positivists and empiricists urged that the activity "science" be defined entirely in terms of sense observation and logical argument—a movement that grew out of courageous opposition to speculative, ungrounded, and metaphysical deadwood that was thought to infest the public science of their time.

Nevertheless, this two-dimensional view of science also has had its costs. It is not true to the behavior and experience of individuals engaged in research. It does nothing to explain why at any given time the choice of problems or the reception of theories may be strikingly different among individuals or like-minded groups who face the same corpus of data. (Examples on this point are the early, quite different responses to relativity in Germany, England, France, and the United States of America.) It also overlooks both the positive, motivating, and emancipatory potential of certain presuppositions, as well as the negative and enslaving role that sometimes has led promising scientists into disastrous error. Einstein and Niels Bohr were rather well matched in navigating the two-dimensional plane of science, as were Schrödinger and Heisenberg. Yet there were among them fundamental antagonisms in terms of programs, tastes, and beliefs, with occasional passionate outbursts among the opponents. The thematic differences, which are at the core of such controversies, do sometimes break through—and they shatter the two-dimensional model.

Above all, the limited view does not explain what a historian, looking at laboratory notebooks or early drafts of distinguished scientists, can sometimes see with stark clarity: the willingness of the scientists to adopt what can only be called a suspension of disbelief about the possible falsification of their hypotheses that emerges from the data before them. Thus while the planar view is satisfactory for understanding the scientists' efforts to further advances in their field inductively, that view requires an amendment to help those who try to understand the original sources and pathways of the creative process. The various thematic[6] propositions can be found to persist for a long time in the individual case, as well as throughout long periods of history. Many themata are

---

6. *Themata* (singular, *thema*) corresponds to the Greek for "that which is laid down"; "proposition"; "primary word."

widely shared, and in a science such as physics they are and have been surprisingly few in number.

The Third Axis of Science

On this view—to elaborate the mnemonic device—a scientific statement is no longer, as it were, an element of area in the two-dimensional plane, but a volume element, an entity in three-dimensional space, with components along each of the three orthogonal (phenomenic, analytic, and thematic) axes. The statements of two scientists are therefore like two volume elements that may not completely overlap and so may have differences in their projections, especially along the z axis. Thematic analysis, then, can serve to identify the particular map of the various themata that, like fingerprints, characterize an individual scientist or part of the scientific community at a given time. For Einstein one can discern the themata supporting his theory construction throughout most of his long scientific career—see, in the References listed at the end of this chapter, especially book 2 *(Thematic Origins)*, chapters 6–9; book 4, chapters 2–4; book 5, chapter 3; and book 6, chapters 7 and 9.

Among other major scientists whose work has lent itself to better understanding through thematic analysis are Galileo (book 6, chapter 4), Thomas Young (book 6, chapter 4), Ernst Mach (book 2, chapter 7; book 5, chapters 1 and 2), A. A. Michelson (book 2, chapter 8), R. A. Millikan (book 3, chapter 2), Heisenberg (book 4, chapter 7), Erwin Schrödinger (book 2, chapter 4), J. Robert Oppenheimer (book 4, chapter 7), Max Planck (book 5, chapter 3), and Steven Weinberg (book 3, chapter 1). In other publications I have discussed other scientists. Thematic analysis in teaching the history of science is shown in the fifth chapter of book 6 to be one of the conceptual tools needed for a full understanding of an event or a case in the history of science.

These studies have shown that themata embraced by opposing scientists often appear in opposing dyads, symbolized by Θ/antiΘ. Examples are: continuum (e.g., in field) versus discontinuum (e.g., in atomism); complexity/simplicity; reductionism/holism; unity/hierarchical levels; causality/probabilism; analysis/synthesis. There are also a few triads, such as evolution/steady state/devolution, or mechanistic/materialistic/ mathematical models. While I have studied primarily themata in physical science, the same findings appear to be applicable also to the other sciences. A list of those found so far can be constructed from the indexes of the books listed at the end, under "thema" and "themata."

The thematic elements of scientific thought become visible most strikingly during a conflict between individuals or groups committed to opposing themata, or within the developing work of a scientist holding on to a thematic concept for as long as possible, until forced to switch to the anti-thema (as in the case of Millikan, on the photoelectric effect; or in the case of Max Planck and the quantum). It is impressive that research has led to the finding that only a relatively small number of themata and anti-themata—perhaps of the order of one hundred—have sufficed throughout modern science. The contrary themata of Heraclitus and Parmenides are still in use. This is one reason to avoid the word *theme* in place of *thema*, for as in music there is no limit to the number of possible themes, there is no longevity nor any overarching generality that can be drawn from them for science.

Additions—such as the introduction of the thematic notion of complementarity in the 1920s—are rare. To be sure, a scientific concept such as "atom" has changed over and over again, from Democritus to this day. But what has not changed is the thematic concept of *discreteness* underlying atomism, which expresses itself in the same way in the ever-changing notion of "atom." That, in brief, is one of the reasons I am not convinced of the theory of the "incommensurability" of theories.

At the same time it also should be pointed out that some scientists function very well without allegiance to a set of thematic ideas, while others are led into error by holding fiercely to an inappropriate thematic idea (Mach, book 5, chapter 2; Felix Ehrenhaft, book 3, chapter 2). Of course not all themata are meritorious. As Francis Bacon warned in discussing the four Idols that can trap the scientific mind, some have turned out to divert or slow the growth of science. Nor have all sciences benefited equally; the holistic viewpoint introduced early in the nineteenth century had advantages in physics but was on the whole a handicap for biology. Also, it should not be necessary to stress that thematic analysis, which arose out of an *empirical study* of actual scientific work, is not an ideology, a school of metaphysics, a plea for irrationality, an attack on the undoubted effectiveness of empirical data and experimentation, or a means for teaching scientists how to do their job better.

Thematic Diversity and Change

We can turn now to one of the puzzles facing every scientist and historian of science. If, as Einstein and others have claimed, the concepts of science are free inventions of the human mind, should that not allow an infinite set of possible thematic axiom systems clamoring for use in constructing a theory to which one's mind could leap or cleave? Virtually every one of these systems would ordinarily be useless for fashioning a theory to encompass the phenomena being studied. How then could there be any hope of success except by chance? The answer must be that the license implied in the leap to an axiom system by the freely inventing mind is the freedom to make a leap, but not the freedom to make any arbitrary leap whatever. The choices available are narrowly circumscribed by a scientist's particular set of themata that filters, constrains, and shapes the style, direction, and rate of advance on novel ground. (See book 4, chapter 2.) And insofar as the individuals' sets of themata overlap, the progress of the scientific community as a group is similarly constrained or directed. Otherwise, the inherently anarchic connotations of "freedom" could indeed disperse the total effort.

Since science is ever unfinished, what is functional in a given field may reveal itself not to be so in the future, and hence there may be a flux of thematic allegiance by the community. The analysis of thematic elements in theories is descriptive, not prescriptive. It may turn out, for example, that the powerfully motivating quest for a general synthesis, so successful from Oersted to Maxwell and from Einstein to our day, could be a trap, as Isaiah Berlin warned in his book *Concepts and Categories*. Himself a dedicated pluralist, Berlin christened the drive toward any grand synthesis the "Ionian Fallacy," so designating the search, from Aristotle to our own day, for the ultimate constituents of the world in some nonempirical sense.

Superficially, the seekers of unified physics, particularly in their monistic exhortations, may seem to have risked falling into that trap—from Copernicus, who confessed that the chief point of his work was to perceive nothing less than "the form of the world and the certain commensurability of its parts," to Max Planck, who exclaimed in 1915 that "physical research cannot rest so long as mechanics and electrodynamics have not been welded together with thermodynamics and heat radiation," to today's string theorists in physics who seem to be offspring of the founding father of science among the ancient Greeks, Thales himself, in their insistence that one entity will explain all. But the scientific profession as a whole has been rescued over the years from becoming mired in traps as might be caused by devotion to a single thema. In practice there is at any time enough variety of commitments—for example, in this case, the existence of a group which in recent years was willing to settle for a pluralistic physics, with a hierarchical set of levels (see some of the publications by Victor Weisskopf and Philip Anderson).

Diversity in the spectrum of themata held by individuals at a given time, with overlap among these finite sets of themata (rather than "anything goes")—that is a formula to explain why the preoccupation with trying to achieve, say, a unified world picture did not lead science to a totalitarian disaster, as an Ionian Fallacy by itself could well have done, nor to an anarchic dispersal of the efforts of the community, nor to a "random walk" without progress, as the paradigm theory of science held at least initially.

At every step, each of the various scientific world pictures in use is generally considered a preliminary version, a premonition of the Holy Grail. Moreover, each of these various, hopeful, but incomplete world pictures that guide scientists at a given time is not a seamless, unresolvable entity. Nor is it usually completely shared even within a given subgroup. Each member of the group is apt to operate with a specific *set* of separable themata. Einstein and Bohr agreed far more than they differed, even though they had profound thematic incompatibilities. Also, most of the thematic currents at any time are not newly minted but adapted from predecessor versions of the world picture, just as many of them will later be incorporated in subsequent versions, as they evolve further.

With this model of the role played by thematic components in the advancement of science, we can understand why scientists need not hold substantially the same set of beliefs to communicate meaningfully with one another, in either agreement or disagreement, while they contribute to the cumulative, generally evolutionary improvement of the state of a science. Scientists' beliefs have considerable fine structure; and within that structure there is room both for thematic overlap and agreement, which generally have a stabilizing effect, and for intellectual freedom, which may be expressed as thematic disagreements.

Innovations emerging from even "far-reaching changes," as Einstein termed the contributions of Faraday, Maxwell, and Hertz, very rarely require from the individual scientist or from the scientific community the kind of radical and sudden reorientation implied in such terms as *Gestalt switch, revolution,* or *discontinuity.* On the contrary, by far most innovations are coherent with a model of evolutionary scientific progress to which most scientists explicitly adhere and which emerges also from the actual historical study of their work. Not being constrained to the two-dimensional plane alone, scientists engage in an enterprise whose saving pluralism resides in its many internal degrees of freedom. What does save science from falling victim to inappropriate presuppositions held for long are the chastening roles both of the coordination with experiment and of the multiple cross-checks of any finding by other scientists who themselves may have started with quite different presuppositions. Thus we can understand why scientific progress is often disorderly, but not catastrophic; why there are many errors and delusions, but not one great fallacy; and how mere human beings, confronting the seemingly endless, interlocking puzzles of the universe, can advance at all in the task of understanding it.

Themata in Science and Elsewhere

Among the costs of the old, two-dimensional definition of "science" has long been a popular image of that pursuit as a cold and lifeless imposition of an authoritarian, dogmatic excess of rationalism. In that view there is no room for the creative play of the intuition or personal preferences (see book 6, chapters 1 and 2). When one adds the role of the thematic dimension in actual research and in the acceptance and rejection of theories, however, this image is shown to be quite false. The corrected version also explains the exaltations of science, from Kepler's ecstasy to the response to Schrödinger's physics by those who shared his thematic propensity for continua ("a fulfillment of that long-baffled and insuppressible desire," as K. K. Darrow put it)—and their opposites, the expressions of dislike and despair that pepper the confessions of scientists. Thus Werner Heisenberg, a master of discontinuum concepts, wrote famously, "The more I ponder the physical part of Schrödinger's theory, the more disgusting it appears to me."

Here we touch also on the fact that the aesthetic and motivational value of themata in science shows that the role of such themata is not so different from the guiding presuppositions and framing worldviews expressed in other creative activities, from the arts to politics. More than that, some themata in a particular science are exemplifications of the same fundamental themata in other sciences, or even in cultural productions far from that of the sciences as such. Take Niels Bohr's complementarity principle as an example. As Bohr himself explicitly noted, he saw complementarity in physics as an expression of one general thema in that relatively small pool of themata from which the human imagination draws for all its endeavors. It is not the case that such an expression is in some way a pale reflection or vague analogy of a principle that is basic only in quantum physics. On the contrary, the situation in quantum physics is a reflection of an all-pervasive principle. Whatever the most prominent factors were which contributed to Bohr's original formulation of the complementarity point of view in physics—whether his physical research, or thoughts on psychology, or his reading in philosophical problems, or the controversy he witnessed between rival schools in biology, or the complementarity demands of love and justice in everyday dealings, to which he often referred—it was the *universal* significance of the role of complementarity which Bohr came to emphasize, to the very end—indeed, even in an interview one day before his death.

Generalizing the case, we may say that each special statement of a certain thema is an aspect of its more general conception. Thus a general thema Θ would take on a specific form in physics that may be symbolized by Θ_ϕ, in studies on mythology and folklore by Θ_μ, and so on. The general thema of discontinuity or discreteness, for example, thus appears in physics as the Θ_ϕ in discussions on atomism, whereas in psychological studies it appears as the thema Θ_ψ of individualized identity. Similarly, the multiple appearances of the Θ of evolution is evident, in different guises, in fields ranging from biology to cosmology, from embryology to economics. One may therefore express any given general Θ as the sum of its specific exemplifications. Hence it is reasonable to expect that thematic analyses, such as those described in the publications cited at the end of this chapter, would be found useful also in a large variety of other fields of study. And in fact this is what has been happening, as I noted in the Postscript to the 1988 edition of *Thematic Origins*.[7]

For pedagogic purposes, the scientific content of curricula has been arranged largely along thematic lines in two national projects, one from the American Association for the Advancement of Science: Project 2061, *Benchmarks for Science Literacy* (New York: Oxford University Press, 1993), and the other in F. J. Rutherford and A. Ahlgren, *Science for All Americans* (New York: Oxford University Press, 1990).

The wide range of these fields bears witness to the fact that there is a much greater commonality of intellectual and motivational resources across areas than is generally acknowledged. The frequently asserted dichotomy between science and humanistic scholarship or productions, while real at many levels, is far less convincing if one looks carefully at the usefulness of the thematic materials in different fields.

Categories of Thought, Kantian and Otherwise

Finally, a word about other concepts occasionally confused with thematic propositions. Examples are metaphors and related notions such as mental models, frames, and schemata. The metaphoric imagination is a lively component of science, alongside others, such as the visual and the thematic (see particularly book 2, chapter 9, and chapter 4 in book 6). But as even its original Greek definition implied, metaphor, unlike themata, serves the traditional function of making conceptual connections between selected similarities; moreover, they are in principle infinite in number.

As noted earlier, another potential confusion with themata involves the concept of "paradigm." But the latter, as usually presented, refers primarily to a social phenomenon in the scientific profession; the para-

---

7. Encyclopedia entries on thematic analysis include Allen Kent (ed.), *Encyclopedia of Library and Information Science*, vol. 61, Supplement 6 (New York: Marcel Dekker, 1987, pp. 332–339); the entry for "*Thema* (Themata)" in the *Dictionnaire d'Histoire et Philosophie des Sciences* (Paris: Presses Universitaires de France, 1999); and the brief survey in J. L. Heilbron, ed., *The Oxford Companion to the History of Modern Science* (Oxford: Oxford University Press, 2003, p. 741).

digm, it is said, does not generally come fully into being until exemplified by the supposedly pervasive acceptance of a particular framework of thought in a speciality. We are told that sooner or later its time is up, however, and another of a potentially infinite number of other paradigms holds sway, until in its turn it, too, is eventually removed in some discontinuous, "revolutionary" development. (As for one of the main arguments made for that idea—i.e., the presumed revolutionary character of relativity theory—Einstein himself always rejected being thought a scientific revolutionary in this field.) By contrast, a thema is found in individual work, as part of an individual's spectrum of themata that no one else may have accepted *in toto;* also, as indicated, themata are found to be finite in number and generally of long duration, and so accentuate the longevity and evolutionary nature of scientific advance.

It may not be necessary to stress here that it is inappropriate to associate themata with Platonic or Jungian archetypes, not least since those, unlike thematic analysis, did not arise from an empirical study. However, there may be a link between individual *research styles* and the embrace of a particular set of themata. Thus Kurt Lewin identified Aristotelian versus Galilean individual modes of thought and showed their persistence in contemporary scientific work. A. C. Crombie, in *Styles of Scientific Thinking in the European Tradition* (London: Duckworth, 1994), presented analogous, interesting case studies on six styles (postulational, experimental, hypothetical, taxonomic, probabilistic and statistical, and historical or genetic).

Among the concepts that may be confused with themata, the most obvious is what Immanuel Kant, following Aristotle, called "Categories." Examples he gave in his *Critique of Pure Reason* involved unity/plurality/totality; possibility/impossibility; existence/nonexistence; necessity/contingency. Apart from other obvious differences, Kant's "Categories" were, as he insisted, to be accepted as "pure concepts of the understanding which apply *a priori* to objects of intuition in general."

Einstein agreed that the mind uses something that might be called categories or schemes of thought, in order "to find our way in the world of immediate sensations." Regarding them as necessary presuppositions for every kind of thinking about the physical world, he stated curtly, "Thinking without the positing of categories and concepts in general would be as impossible as is breathing in a vacuum."[8] But Einstein also insisted on an essential difference, showing that those categories are not Kant's concepts but instead are quite close to thematic propositions. He insisted that his categories are not unalterably *a priori*, conditioned by the very nature of our mind; rather, they arise from and are subject to change by the unfettered imagination; they are "free conventions," justified only by their usefulness.[9] As he puts it, the selection of "'categories' or schemes of thought . . . is in principle entirely open to us, and whose qualification can only be judged by the degree to which their use contributes to making the totality of the content of consciousness 'intelligible.'" Therefore, it is possible that, far from being frozen into any position *a priori*, some scientists have, as noted, changed their thematic allegiances dramatically. (For example, Planck turned from an early Machian to an opponent, and Wilhelm Ostwald first rejected and then accepted atomism in chemistry.)

We end by noting an unsolved puzzle: what is the source of a person's particular set of thematic concepts and hypotheses? It is possible that the origin of themata in individual cases will someday be approached through studies of the nature of perception and apperception—the psychodynamics of the development of concepts in early life. But for our part, the task continues to be the study of the role played by recurring general themata among individual scientists and the profession as a whole.

---

8. Einstein, "Reply to Criticisms," in Paul A. Schilpp (ed.), *Albert Einstein, Philosopher-Scientist* (Evanston, IL: Library of Living Philosophers, 1949), pp. 673–674, 937–940.

9. Einstein was undoubtedly aware of some of the long history of the conventionality concept, at least in the form it appeared in works we know he had read, such as Henri Poincaré's books and the 1907 essay, "Experience and the Law of Causality," by Philipp Frank.

References

1. Holton, Gerald. "Über die Hypothesen, welche der Naturwissenschaft zu Grunde liegen." Pp. 351–425 in Adolf Portmann, ed., *Eranos-Jahrbuch XXXI/1962, Der Mensch, Führer oder Geführter im Leben*. Zurich: Rhein-Verlag, 1963.
2. ——— *Thematic Origins of Scientific Thought: Kepler to Einstein*. Cambridge: Harvard University Press, 1973; rev. ed., 1988.
3. ——— *The Scientific Imagination: Case Studies*. Cambridge: Harvard University Press, 1998.
4. ——— *The Advancement of Science, and Its Burdens*. Cambridge: Harvard University Press, 1998.
5. ——— *Science and Anti-Science*. Cambridge: Harvard University Press, 1993.
6. ——— *Einstein, History, and Other Passions*. Cambridge: Harvard University Press, 2000.

— 11 —

The Imperative for Basic Science That Serves National Needs

The public attitude toward science is still largely positive in the United States.[1] For a vocal minority, however, the fear of risks and even catastrophes supposedly resulting from scientific progress has become paramount. Additionally, in what has been called the "science wars," central claims of scientific epistemology have come under attack by academics outside science, and that demeaning of science has even helped some politicians to disregard scientific findings altogether.[2] Some portions of the political sector consider basic scientific research far less worthy of governmental support than applied research, even while other politicians have castigated the support of applied research as "corporate welfare." And Senator Harry Reid has warned that all too many are now beginning "to see science as a luxury that can be reduced or abandoned."

Amidst the choir of dissonant voices, Congress has been concerned to determine what is being called "a new contract between science and society" for the post–Cold War era. As the late Representative George E. Brown, Jr., stated, "A new science policy should articulate the public's interest in supporting science—the goals and values the public should expect of the scientific enterprise."[3] Whatever the outcome, the way science has been supported during the past decades, along with the priori-

---

1. I wish to acknowledge the essential part played by Dr. Gerhard Sonnert in composing this chapter.

2. Bruno Latour, in *Critical Inquiry* 30, no. 2 (2004). See also Steven Weinberg, *Facing Up* (Cambridge: Harvard University Press, 2001).

3. Along the same lines, see the "Ehlers Report": House Committee on Science, *Unlocking Our Future: Toward a New National Science Policy* (Washington, DC: September 1998).

ties and the motivation for such support, are undergoing changes, with consequences that may well test the high standing that American science has achieved over the past half century.

Reevaluating the Covenant between Science and Society

In this climate of widespread soul-searching, my aim is to propose an imperative for an invigorated science policy that adds to the well-established arguments for government-sponsored basic scientific research. In a novel way, that imperative couples basic research tightly with the national interest. This chapter presents, in outline, a conceptual framework for such a science policy.[4]

The seemingly quite opposite two main types of science research projects that have been vying for support in the past and to this day are often called basic or "curiosity driven" versus applied or "mission oriented." While these common characterizations have some usefulness, they harbor two crucial flaws. The first is that in actual practice these two contenders usually interact and collaborate closely; they are not clear-cut antitheses or inherently opposed in claims for support, despite what the most fervent advocates of either type may think. The history of science clearly teaches that many of the great discoveries that ultimately turned out to have beneficial effects for society were motivated by pure curiosity without thought for such benefits; equally, the history of technology recounts magnificent achievements in basic science by those who embarked on their work with practical or developmental interests.

As the scientist-statesman Harvey Brooks appropriately commented, we should really be talking about a "seamless web."[5] The historian's eye perceives the seemingly initially unrelated pursuits of basic knowledge, technology, or instrument-oriented developments in today's practice to be the weaving of a single, tightly woven fabric. Harold Varmus, the former director of the National Institutes of Health (NIH), eloquently

---

4. Lewis Branscomb, in "The False Dichotomy: Scientific Creativity and Utility," *Issues*, 16, no. 1 (Fall 1999): pp. 66–72, provides some of the historical background and discusses policy issues involved in this initiative.

5. Harvey Brooks, "The Changing Structure of the U.S. Research System," in H. Brooks and R. Schmitt, *Current Science and Technology Issues: Two Perspectives* (Washington, DC: George Washington University, 1995).

acknowledged the close association of the more "applied" biomedical advances with progress in the more "basic" sciences: "Most of the revolutionary changes that have occurred in biology and medicine are rooted in new methods. Those, in turn, are usually rooted in fundamental discoveries in many different fields. Some of these are so obvious that we lose sight of them—like the role of nuclear physics in producing radioisotopes essential for most of modern medicine." Varmus went on to cite a host of other examples that outline the seamless web between medicine and a wide range of "basic" science disciplines.

The second important flaw in the usual view is that these two widespread and ancient modes of thinking about science, pure versus applied, have tended to displace and derogate another, third way—I have called it "Jeffersonian science." This third mode now deserves the attention of researchers and policy makers. But I shall by no means advocate that Jeffersonian science replace the other two modes. Science policy should never withdraw from either basic or applied science. I argue that the *addition* of a third mode, one that can be called the Jeffersonian mode, to an integrated framework of science policy would tremendously contribute to mobilizing widespread support for science and to propelling both scientific and societal progress. Before I turn to a discussion of it, we should briefly survey the other two modes of scientific research.

The Newtonian and Baconian Modes of Scientific Research

The concept of pursuing scientific knowledge "for its own sake," letting oneself be guided chiefly by the sometimes overpowering inner necessity to follow one's curiosity, has been associated with the names of many of the greatest scientists, but most often with that of Isaac Newton. His *Principia* (1687) may well be said to have given the seventeenth-century Scientific Revolution its strongest forward thrust. It can be seen as the work of a scientist motivated by the abstract goal to achieve eventually complete intellectual "mastery of the world of sensations" (Max Planck's phrase). Newton's program has been identified with the search for *omniscience* concerning the world accessible to experience and experiment, and hence with the primary aim of developing a scientific world picture within which all parts of science cohere. In other words, it is motivated by a desire for better and more comprehensive scientific knowledge. That approach to science can be called the *Newtonian mode*. In all past and present activity in the Newtonian mode, the hope for practical and benign *applications* of the knowledge gained in this way is a real but secondary consideration.

The second of the main styles of scientific research is popularly identified with "mission-oriented," "applied," or "problem-solving" science. Here we find ourselves not among those who, in the Newtonian mode, were essentially in pursuit of omniscience, but those who might be said to follow the call of Francis Bacon. Bacon urged the use of science not only for "knowledge of causes and secret motion of things," but also in the service of *omnipotence*, "the enlarging of the bounds of human empire, to the effecting of all things possible." Even stripped of its overarching and over-extended rhetoric, the latter aspect of this approach is characteristic not of those whose search proceeds without prime regard for applications, but of those to whom the "effecting of all things possible" is the main prize.

Such research in what might be called the *Baconian mode* has been carried out more likely in the laboratories of industry than of academe. Unlike basic research, mission-oriented research by definition hopes for practical, and preferably rapid, benefits; and it proceeds to produce applications where it can, by using basic knowledge already known.

The Jeffersonian Mode of Scientific Research

There is a mode of research which does not quite fall under the headings discussed so far—and it may open a new window of opportunity in the current reconsiderations of what kinds of science are worth supporting, not least in Congress and the federal agencies. It is a conscious combination of aspects of both previously noted modes, and it is best characterized by the following formulation: *The specific basic research project is motivated by and placed in an area of basic scientific ignorance that is thought to lie at the heart of a social problem.* Its main goal is to remove that basic ignorance in an uncharted area of science, and thereby to attain knowledge which will have a fair probability—even if it is years distant—of being brought to bear on a persistent, debilitating national (or international) problem.

An early and impressive example of this sort was Thomas Jefferson's decision to launch the Lewis and Clark expedition into the western parts of the North American continent. Jefferson, who declared himself most happy when engaged in some scientific pursuit, correctly understood that the expedition would serve basic science, by bringing back maps, samples of the unknown fauna and flora, and observations on the native inhabitants of that blank area on the map. At the same time, however, Jefferson realized that knowledge about the North American continent would eventually be desperately needed for such practical purposes as establishing relations with the indigenous peoples, and furthering the eventual westward expansion of the burgeoning United States population. The expedition thus implied a *dual-purpose style of research:* basic scientific study of the best sort (suitable for an academic Ph.D. thesis, in modern terms) with no sure, short-time "payoff"—but targeted in an area where there is a recognized problem affecting society. I have called this style of basic scientific research the Jeffersonian mode.[6]

This third mode of defining the site for scientific research, which I am proposing here for our own period, offers a way to avoid the dichotomy of Newtonian versus Baconian styles of research while supplementing both. In the process it can make public support of all types of research more palatable to policy makers and taxpayers alike. It is after all not too hard to imagine basic research projects that hold the key to alleviating well-known societal dysfunctions. Even the "purest" scientist is likely to agree that much remains to be done—to give only a few examples—in cognitive psychology, the biophysics and biochemistry involved in the process of conception, the neurophysiology of the senses such as hearing and sight, molecular transport across membranes, or the physics of nano-dimensional structures, to name a few. The result of

---

6. I have presented previous treatments of this concept in the following publications: The Jefferson Lecture of 1981, in *The Advancement of Science, and Its Burdens* (Cambridge: Harvard University Press, 1998); in "What Kinds of Science Are Worth Supporting? A New Look, and a New Mode," in *The Great Ideas Today* (Chicago: Encyclopaedia Britannica, 1988), pp. 106–136; chap. 4 in *Science and Anti-Science* (Cambridge: Harvard University Press, 1993); and pp. 117–127 in L. Branscomb, G. Holton, and G. Sonnert, *Science for Society: Cutting-Edge Basic Research in the Service of Public Objectives* (Cambridge, MA: Kennedy School of Government, Harvard University, 2001).

such basic work, one could plausibly expect, will give us in time a better grasp of complex social tasks such as, respectively, childhood education, family planning, improving the quality of life for handicapped persons, the design of food plants that can use inexpensive (brackish) water, and improved communication devices.

Other research sites for the Jeffersonian mode include, for example, the physical chemistry of the stratosphere; the complex and interdisciplinary study of global changes in climate and in biological diversity; that part of the theory of solid state which makes the more efficient working of photo-voltaic cells still a puzzle; bacterial nitrogen fixation, and the search for symbionts that might work with plants other than legumes; the mathematics of risk calculation for complex structures; the physiological processes governing the aging cell; the sociology underlying the anxiety of some parts of the population about mathematics, technology, and science itself; or the anthropology and psychology of ancient tribal behavior that appears to persist to this day and may be at the base of genocide, racism, and war in our time.

It is of course true that Jeffersonian-type arguments are already being made from time to time and from case to case; that is, problems of practical importance are used to justify federal support of basic science. For instance, research sponsored by the National Science Foundation (NSF) in atmospheric chemistry and climate modeling is linked to the issue of global warming, and Department of Energy support for plasma science is justified as providing the basis for controlled fusion. The National Institutes of Health (NIH) have been particularly successful in supporting Jeffersonian-type efforts in health-related basic research. Yet what seems to be missing is an overarching theoretical rationale and institutional legitimation of Jeffersonian science, both within the federal research structure and in academe.

The current interest in rethinking science and technology policy beyond the confining dichotomy of basic versus applied research has spawned some efforts kindred to ours. In Donald Stokes's framework, the linkage of basic research and the national interest appeared in what he called "Pasteur's Quadrant,"[7] which overlaps to a degree with what

---

7. D. E. Stokes, *Pasteur's Quadrant: Basic Science and Technological Innovation* (Washington, DC: Brookings Institution Press, 1997).

I had termed earlier the Jeffersonian mode. My approach also heeds Lewis Branscomb's warning that the degree of utility considerations in motivating research does not automatically determine the nature and fundamentality of the research carried out. Branscomb appropriately distinguishes two somewhat independent dimensions of *how* and *why*: the character of the research process itself (ranging from basic to problem-solving) and the motivation of the research sponsor (ranging from knowledge-seeking to concrete benefits). For instance, a basic research process—which for Branscomb comprises "intensely intellectual and creative activities with uncertain outcomes and risks, performed in laboratories where the researchers have a lot of freedom to explore and learn"—may characterize research projects with no specific expectations of any practical applications, as well as projects that are clearly intended toward application. Branscomb's category of research that is both motivated by practical needs and conducted as basic research is coherent with my concept of Jeffersonian science.[8]

I can summarize the theoretical framework in the form of a diagram representing Newtonian, Baconian, and Jeffersonian science:

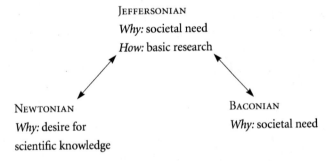

The Carter-Press Initiative

Jeffersonian science is not an empty dream. Related science policy initiatives exist. Here I briefly turn to a concrete example of an attempt to institute a Jeffersonian research program on a large scale. Long neglected, that effort is eminently worth remembering as the covenant between science and society is reevaluated. In what follows, I can of-

---

8. See Branscomb, "False Dichotomy."

fer only a short mention of what may be called the Carter-Press initiative.[9]

In November 1977, upon President Carter's request, Frank Press, presidential science adviser and director of the Office of Science and Technology Policy, polled the federal agencies about *basic* research questions whose solutions, in the view of these agencies, were expected to help the federal government significantly in fulfilling its mission. The resulting master list, which was assembled in early 1978, turned out to be a remarkable collection of about eighty research questions considered by the heads of federal agencies (including the Departments of Agriculture, Defense, Energy, State, and NASA) to be good science—good, here, in the sense of having expected practical pay-offs—but also to resonate with the intrinsic standards of the scientific community.[10]

Consider, for instance, a research question from the Department of Agriculture: "What are mechanisms within body cells which provide immunity to disease? Research on how cell-mediated immunity strengthens and relates to other known mechanisms is needed to more adequately protect humans and animals from disease." That question, framed in 1978 as a basic research question, was to become a life-and-death issue for millions only a few years later, with the onset of the AIDS epidemic. The selection of this research topic illustrates that Frank Press's "Jeffersonian" initiative was able in advance to target a basic research issue whose potential benefits were understood in principle at the time, but whose dramatic magnitude could not have been foreseen (and might well not have been targeted in a narrowly application-oriented research program).

Other remarkable Jeffersonian research questions included one by the Department of Energy about the effects of atmospheric carbon dioxide concentrations on the climate and on global social, economic, and political structures; and one by the Department of Defense about

---

9. For details, see Gerhard Sonnert, *Ivory Bridges: Connecting Science and Society* (Cambridge: MIT Press, 2001).

10. It should be added here that the agency heads were in the position to make meaningful scientific suggestions thanks in good part to two of Frank Press's predecessors, Science Advisers Jerome Wiesner and George Kistiakowsky. They had helped to build serious science research capacities into the various federal agencies, thus assuring that highly competent advice was available from staff scientists within the agencies.

superconductivity at higher temperatures—almost a decade before the sensational breakthrough in this area.

Revitalizing the Jeffersonian Research Agenda

The Carter-Press initiative quickly slid into oblivion with the end of Carter's presidency; yet it does not deserve to be forgotten. A revitalization of the Jeffersonian mode of science would provide a promising additional model for future science policies, one that would be especially relevant in the current state of disorientation about the role of science in society. I conclude this chapter with three general observations.

For many scientists, the institution of funding for research in the Jeffersonian mode would be liberating. Some would be glad to target their work in a way that makes beneficial practical outcomes for society likely. Others who intend to do basic research in this mode, in the defined areas of national interest, would be shielded from pressures to demonstrate the *immediate* social usefulness of their specific projects in their grant applications. Once these areas of interest are determined, the decisions on awards of research grants can proceed according to the usual standards of merit.

Moreover, a Jeffersonian agenda provides an overarching rationale for governmental support of basic research that is both theoretically sound and easily understood by the public. It defuses the increasing charges that science is not sufficiently concerned with "useful" applications, for that third mode of research is precisely located in the area where the national—and international—welfare is a main concern. The way "basic" research in the interest of health is already legitimized and supported under the auspices of the National Institutes of Health may well serve as a successful exemplar that other sciences could adopt.

Finally, the strengthened public support for science, induced by a visible and explicit Jeffersonian agenda, is likely to generalize and transfer to other sectors of federal science policy. (Again, I do not advocate the total replacement of the Newtonian and Baconian modes by the Jeffersonian mode; all must be part of an integrated federal science policy.) Even abstract-minded high-energy physicists have learned the hard way that their funding depends on a generally favorable attitude to science as a whole. Moreover, they too can be proud of the use of the campus cyclotron for cancer treatment and the production of radioisotopes, the use of nuclear magnetic resonance (NMR) or synchrotrons for imaging, and more. Nor should we forget the valued participation of pure theorists on important government science advisory panels—nor their sudden usefulness, with historic consequences, during both World War I and World War II.

From every perspective, ranging from the purely cultural role of science to national preparedness, even the "purest" scientists can continue properly to claim their share of the total support given to basic science. But that total sum can more easily be enlarged by the change I advocate in the public perception of what basic research can do for the pressing needs of humankind.

— 12 —

The Rise of Postmodernisms and the "End of Science"

In his remarkable essay, "The Apotheosis of the Romantic Will,"[1] Isaiah Berlin articulates a key question facing contemporary historians of ideas. He begins with the observation that beliefs have entered our culture that "draw their plausibility" from a deep and radical revolt against the central tradition of Western thought. That central tradition rested on the "pillars of the social optimism," which had found its fullest expression in the Enlightenment, "that the central problems of men are, in the end, the same throughout history; that they are in principle solvable; and that the solutions form a harmonious whole."

Berlin notes that these pillars "came under attack toward the end of the eighteenth century by a movement first known in Germany as *Sturm und Drang,* and later in the many varieties of romanticism . . . and the many contemporary forms of irrationalism of both the right and the left, familiar to everyone today." In our time, in the alleged absence of "objective rules," the new rules are those made up by the rebels: "Ends are not . . . objective values," and "ends are not discovered at all but made, not found but created." He concludes: "The prophets of the nineteenth century predicted many things . . . but what none of them, so far as I know, predicted was that the last third of the twentieth century would be dominated by . . . the enthronement of the will of individuals or classes, and the rejection of reason and order as being prison houses of the spirit. *How did this begin?*"[2]

---

1. In Isaiah Berlin, *The Crooked Timber of Humanity: Chapters in the History of Ideas,* ed. Henry Hardy (New York: Vintage Books, 1992).

2. Ibid., pp. 208–213

As if to ensure that this question be considered central to the understanding of our age, he adds that the explosion of irrationalism is one of the "outstanding characteristics of our century, the most demanding of explanation and analysis."[3] Elsewhere he also appeals to seek the causes of "what appears to me to be the greatest transformation of Western consciousness, certainly in our time."[4]

As this essay focuses chiefly on the aspects concerning science, let me rephrase Berlin's question: How did it come about that, in the phrase coined by the philosopher Susan Haack, we have passed again in many areas into an "Age of Preposterism";[5] that, for example, scientists, who are now in a period of spectacular advances of knowledge across the board, find a whole array of highly placed academics and journalists asserting that their (never fully achieved but essentially motivating) hopes to reach objective truths are in vain because there is no difference between the laws scientists find in nature and the arbitrary rules that govern baseball games; that science is "just one language game among others"? Why, at this moment in history, are serious claims being made that we must "abolish the distinction between science and fiction"; that "the natural world has a small or non-existent role in the construction of scientific knowledge"; and that, as the title of a current bestseller has it, we are at *The End of Science: Facing the Limits of Knowledge in the Twilight of the Scientific Age?*[6]

3. Ibid., p. 1.
4. Isaiah Berlin, *The Roots of Romanticism* (Princeton: Princeton University Press, 1999), p. 20.
5. Susan Haack, *Manifesto of a Passionate Moderate* (Chicago: University of Chicago Press, 1998); *Defending Science—Within Reason: Between Scientism and Cynicism* (Amherst, NY: Prometheus Books, 2003).
6. For documentation and examples, see chap. 6 in G. Holton, *Science and Anti-Science* (Cambridge: Harvard University Press, 1993); chap. 1 in G. Holton, *Einstein, History, and Other Passions: The Rebellion against Science at the End of the Twentieth Century* (Cambridge: Harvard University Press, 2000); Perry Anderson, *The Origins of Postmodernity* (London, New York: Verso, 1998); Alan Sokal and Jean Bricmont, *Fashionable Nonsense: Postmodern Philosophers' Abuse of Science* (New York: Picador, 1998); Noretta Koertge, ed., *A House Built on Sand: Exposing Postmodernist Myths about Science* (Oxford, New York: Oxford University Press, 1998); Steven Weinberg, *Facing Up: Science and Its Cultural Adversaries* (Cambridge: Harvard University Press, 2001); and John Horgan, *The End of Science: Facing the Limits of Knowledge in the Twilight of the Scientific Age* (Reading, MA: Addison-Wesley, 1996).

These are only a few examples of a stream of derogations issuing from academe and the media. Happily, though, my concern here is not with the details of the current manifestation of what has been called the war on science, but rather with examples of its historic lineage, with earlier phases of Isaiah Berlin's "Romantic Revolt."[7] Here we must begin our analysis by recognizing that any such movement as Berlin identified is best understood as a reaction against what went before, a reaction against what became so unsatisfactory or even intolerable as to cause the revolt.

The Charges against Modernity

Historically, the most obvious and early reaction of this sort was the response to the breakthroughs in the seventeenth century that formed science and simultaneously signaled a great rupture from the ancient worldview in which the individual, in principle, could be both intellectually *and* spiritually comfortable. As one of the direct ancestors of romanticism, Johann Gottfried Herder, put it, premoderns could still understand and grasp "the solid order of nature, and they lived safely within it."[8] After the rise of modern science, in the words of Jean Paul Richter, mankind found itself lost in a mechanistic solar system, that "all-powerful, blind, lonesome machine."

Each field has its own date for the onset of offensive modernity. For science, the plausible date when "human character changed" is not, as Virginia Woolf put it, "in or about December 1910," but February 10, 1605, when Johannes Kepler, while working on his *Astronomia Nova*, laid out his breathtaking ambition in a letter to his friend, Herwart von Hohenburg:

> I am much occupied with the investigation of the physical causes [of the motions of the solar system]. My aim in this is to show that the celestial machine is to be likened not to a divine organism, but rather to a clockwork..., insofar as nearly all the manifold movements are carried out by means of a single, quite simple... force, as in the case of a

---

7. Berlin, *Crooked Timber*, p. 229.
8. Quoted in Alexander Gode-von Aesch, *Natural Science in German Romanticism* (New York: Columbia University Press, 1941), p. 51.

clockwork [all motions are caused] by a simple weight. Moreover, I shall show how this physical conception is to be presented through calculation and geometry.

This approach, hugely successful on its own terms, was soon regarded by one side as triumphs of reason and experiment but by the other as a deeply felt assault on mankind's stature and self-confidence—as culture shock and epistemic trauma, to use modern argot. The list of indictments was long: Galileo's completion of the de-centering of the human abode in January 1610, when his telescope revealed the existence of moons around Jupiter, thereby launching mankind, with reduced significance, into the unbound Copernican void and causing John Donne's anguished outburst of 1611, "Tis all in peeces, all cohaerence gone"; the elevation of the quantitative in nature over the qualitative and the objective and skeptical over the subjective and mystical; the separation of the natural from the supernatural; the validation of rational over intuitive discourse; and more. Among the results of these changes were the "disenchantment of Nature" (in Max Weber's term), a turning away from the ancient fascination with individual, wondrous, portentous instances toward the search for general and overarching laws, from individual belief toward shareable results; the downgrading of the contemplative relation with nature, in favor of active intervention; and above all the mechanization of the model of the universe in Isaac Newton's published writings—if one neglects the few passages hinting that Newton was no Newtonian but rather privately a life-long searcher for the nature of the divinity. Of course, the change of the predominant worldview was slow and complex, with seemingly contradictory strains co-existing for a long time (as scholars such as Alexandre Koyré, E. A. Burtt, and Hèléne Metzger pointed out long ago). But appropriately, the word *modernity* and its cognates entered the English language starting as early as the 1620s. As the romantic dramatist Heinrich von Kleist put it later: "Paradise is now bolted and barred."

More elements of modernism followed—the postulation of the mind-body dualism, the findings of evolutionism, some of the tenets of psychoanalysis—each adding to the de-divinization of man and Nature, which Friedrich Schiller had called the *Entgötterung der Natur*. As Koyré remarked: "The mighty, energetic God of Newton who actually 'ran' the universe according to His free will and decision, became, in quick succession, a conservative power, an *intelligentia supra-mundana*, a '*Dieu fainéant*'. . . . The infinite Universe of the New Cosmology . . . inherited all the ontological attributes of Divinity. Yet only those—all the others the departed God took away with Him."[9]

During the last two centuries, it has seemed to many that science became ever more arcane, and technology a blind juggernaut. Some overreaching remarks by scientists, from Laplace to Wilhelm Ostwald to Stephen Hawking, did not help either. But all the deeds and misdeeds charged against the mindset of the Enlightenment, against all the excesses of the project of modernity and its inherent practical and psychological incompleteness, must be understood with sympathy. They served periodically to coalesce a critical mass of strenuous objectors of a great variety—from the clergy under Pope Urban VIII in the days of Galileo to the brilliant poets, Keats, Byron, Shelly and Blake (William Blake especially, who regarded Newton as his personal Satan); from the mystic Jakob Böhme to a recent lecturer who, to great applause, called for the "return to the Holy Darkness." Thus, the periodic rebellions against the worldview evolving from the rise of modern science—itself to a high degree the child of a reaction in the seventeenth century against the canon of the ancients—must be understood as episodic upwellings of a bipolar sentiment deeply rooted in the human psyche: one part an aching mourning for a glamorized version of the earlier state of humanity, the other part a desperate longing for a utopian restitution, in new form, of what had been lost. Throughout history, passionate sentiment may dominate for a few decades, even inspire immortal works by philosophers, poets, composers, and artists, and then may largely subside in the face of a slowly rising opposition to its excesses—with a lingering undercurrent which, on the personal scale, each individual may feel but which, on a large scale, prepares for the next rise, the next phase of the Romantic Rebellion.

To illustrate the context and variety of these major outbreaks over the past two centuries, I select here from many worthy examples two works that focus on the intellectual rather than the social factors of the Romantic Rebellion. Furthermore, to signal the widely *differing* aspects of the phenomenon we are discussing here, I have chosen one example

---

9. Alexandre Koyré, *From the Closed World to the Infinite Universe* (New York: Harper and Brothers, 1958), p. 276.

that was relatively benign and one that was diabolically destructive. But both have left traces in today's distinctive versions of the Romantic Rebellion.

From Kant to Romantic Science

The first example is the ascent of *Naturphilosophie*, prominent for a few decades at the beginning of the nineteenth century. One of its main sources is, perhaps surprisingly, a majestic figure in the history of ideas who arguably may be represented as both an admirer of the seventeenth-century scientific revolution and, through idiosyncratic readings of his work, as a rebel against it. I speak of Immanuel Kant of Königsberg, that veritable mountain from which different streams of thought descended, like a peak on the Continental Divide giving birth to rivulets that diverge, grow, and eventually end up in different oceans.

Kant's deep interest in Newtonian science started in his student years. At age thirty-one he published his "Universal Natural History and Theory of the Heavens," with a subtitle ending with the words "Treated according to Newtonian Principles." It is full of remarkable anticipations of subsequent cosmological findings and theories. But importantly—unlike the Newtonians who followed only Newton's mechanistic *Principia* and *Opticks*—the Kant of 1755 retained a role in his analysis for God, who directs Nature after having created space, time, matter, and the laws of Nature. By the time he published the *Critique of Pure Reason* in 1781, however, he despaired of providing a proof for the existence of God. (He left the door open, though, by denying that there ever could be a *disproof* of His existence—a point of major importance to his later pietistic followers.) Also, in the *Critique of Pure Reason* space and time became conditions of human knowledge based in Categories, in intuitions pre-existing in every mind. Thus he launched his transcendental idealism, in which the form of experience is supplied by the human mind while the real material world outside the human self is the source of experience, which comes to us through sensations. The later idealists, who believed themselves to follow Kant, went much further. Thus, Friedrich Wilhelm Schelling would hold that there was no need to dirty one's hands with experiments.

Five years after his first *Critique* came Kant's *Metaphysical Foundations of Natural Science*. In it, a point essential for our purposes, is Kant's view that "motive forces" of only two kinds, attraction and repulsion, provide the fundamental attributes of matter. This theme of two opposing forces determining all natural phenomena had already preoccupied the alchemists and the sixteenth-century iatrochemists such as Paracelsus and Jan van Helmont. As elaborated by Kant, the polarity of forces masks a "hidden [*versteckte*] identity," which allows one to hypothesize a unity, a *"Grundkraft,"* a fundamental force of which all other forces are variants. We recognize here a thematic line that goes back to Thales of Ionia, who looked for one substance or essence to explain all phenomena of the material world, and forward to the attempts of our contemporary physicists to unify the four main forces of nature into one. The old Ionian Enchantment, active at the very beginning of science, also infected Kant, for whom Unity was the first of all his Categories; as we shall see, it also inspired those who regarded themselves as his pupils.

The works of Immanuel Kant provided the wellsprings from which issued two contrary main directions of thought. One is exemplified in the later scientific work of major scientists such as Hermann Helmholtz, Emil Du Bois-Reymond, and Rudolf Virchow; these successors to the Newtonian synthesis embraced the experimental contact with nature and the interest in Newtonian science that had been part of Kant's thought. But on the other side, Kant could be read (or misread, as Friedrich Schlegel did) as the father of a very different, new view of science, one infused with the Romanticism of the "Nature Philosophers."

Those *Naturphilosophen* were numerous enough to create a critical mass of thinkers whose ideas exploded into the intellectual life of the period. They included Friedrich Schelling, philosopher, friend of Hegel; the brothers Friedrich and A. W. Schlegel; Novalis; Goethe; and all their influential followers. Schelling shared with most of his friends the view that nature is an organism rather than a mechanism, that the world contains a single elemental force *(Urkraft)* which, thanks to its inherent polarity, produces a conflict between its diverse exemplifications in nature; that matter, contrary to Newtonian physics, was never inert but was alive and subject to a conflict that explained growth, decay, and chemical reactions. In opposition to the rationalism favored by the Enlightenment, Schelling published books, such as *Philosophie der Natur* (1797), celebrating intuition, and he founded two journals on so-called speculative physics.

Johann Wilhelm Ritter—chemist, physicist, physiologist, a tragic and unruly figure, though a prolific experimenter—dabbled in occultism. Like many others in this group he believed in a World Soul animating nature and causing all phenomena to be interdependent and unified. Lorenz Oken, natural scientist and philosopher, also supported an enthusiastic romanticism through his concepts of the *Ur*-man and the evolution of all life forms from a primal slime. And Christian Weiss, with fundamental contributions to crystallography, was deeply influenced by Kant, by Schelling, and by the philosopher Johann Gottlieb Fichte. Fichte's book, *Wissenschaftslehre* (1794), powerfully infused the Romantic movement with his idea of the primacy of the individual ego, the freedom from objective rules, and the construction of value by the creative human self rather than by reliance on the historic canons.

Despite many differences, these men had much in common. L. Pearce Williams has pointed out[10] that they all were born within a span of eight years in the 1770s, the decade that launched, first in literature, the period known after the play *Sturm und Drang* by F. M. Klinger (1777). They grew up in the aftermath of the Terror and the disintegration of revolutionary France; recoiling from the upheaval, they came to believe that "without God, there could be no social order." They gave free run to their emotions in their writings, often taking excessive risks and always remaining open to the arts, especially to literature. As Williams put it, "All felt the near ecstasy of creativity springing from the active mind. Spirit was as real to them as body. All underwent youthful crises, and discovered Kant as the answer to their personal angst." Kant's loophole for the existence of God (by the impossibility of disproving His existence) was, as they read it, a liberation that gave a place to the highest intelligence in nature, as well as freedom for rampant speculation in science—precisely a point to which Kant would have objected strongly.

Even the necessarily fragmentary characterizations given here indicate persistent themes parallel to some ideas espoused in the current phase of the Romantic movement. But there are also fundamental differences. The most obvious one is that almost every Nature Philosopher of the eighteenth and the first part of the nineteenth century had, from their

---

10. L. Pearce Williams, "Kant, 'Naturphilosophie' and Scientific Method," in *Foundations of Scientific Method: The Nineteenth Century*, ed. R. N. Giere and R. S. Westfall (Bloomington: Indiana University Press, 1973).

perspective, an intense and honorable interest in scientific matters, even if non-Romantic scientists, such as Jöns Jacob Berzelius and Justus von Liebig, would have none of it. Liebig famously cried out: "the activities of the *Naturphilosophen* are the pestilence, the black death of the century." But most scientifically inclined Nature Philosophers tried to, and some did, contribute to science in their way. For example, after William Herschel discovered invisible infrared light at one end of the spectrum, invisible ultraviolet light was discovered to exist at the other end by Johann Wilhelm Ritter, reasoning entirely from the Romantic penchant for analogy and polarity. Another person ironically did stumble on a most fundamental scientific advance by a fanciful interpretation of Kant's ideas. It is a case worth lingering on for a moment. This man, for whom I have a special fondness, was the Danish scientist Hans Christian Oersted—a typical Romantic, in his incessant attempts to re-enchant nature by endowing her with *Geist* or spirit (as he wrote in *The Soul in Nature*), in his pietism, even in his effusion into poetry.[11]

Oersted made room in his own research for speculation and intuition and for what he called, in the happy phrase, the willingness to allow the scientific imagination to be guided by an "anticipating consonance with Nature." On the other hand, arguing that the human mind reflects Divine reason only very dimly, he knew he also had to subject those intuitions eventually to *experiment*. And that he did, preserving in that respect a continuity with the Newtonians, despite the ideological differences and mutual disdain between these two worldviews. He studied in Berlin for a time under Fichte and the brothers Schlegel, but his idiosyncratic reading of Immanuel Kant was Oersted's main guide from the beginning. His doctoral thesis in philosophy of 1799 was on the "Metaphysics of External Nature," a recasting and extension of the book *Metaphysische Anfangsgründe* published by the man whom Oersted called, at the very beginning of the treatise, "the immortal Kant."[12]

---

11. Most of Oersted's scientific papers, with useful introductions, are available in Kristine Meyer, ed., *H. C. Ørsted's Scientific Papers*, 3 vols. (Copenhagen: Andr. Fred. Høst and Søn, 1920); in *Selected Scientific Works of Hans Christian Ørsted*, ed. Karen Jelved, Andrew D. Jackson, and Ole Knudsen (Princeton: Princeton University Press, 1998); and in *H. C. Ørsted and the Romantic Legacy*, ed. Robert Brain and Ole Knudsen (Dordrecht: Kluwer, in press).

12. Jelved et al., *Selected Scientific Works*, p. 80.

Oersted accepted Schelling's *interpretation* of Kant, that nature's phenomena were to be explained by a "conflict" between opposing exemplifications of the unitary force that sustained matter itself. So in an essay[13] published in 1805 at age twenty-eight, Oersted announced that electricity and magnetism—then regarded as completely different and unrelated forces by all mainstream scientists in Europe—were, on the contrary, related "dynamic processes," explainable as "the interaction of opposite fundamental forces in a different form." Seven years later he explicitly used Kantian ideas of the one basic force underlying its polar exemplifications to explain that under different circumstances these processes should take the form of electricity, magnetism, heat, light, and chemical reactions, depending on the experimental conditions.

It took until one evening in April 1820 for Oersted to get around to testing his fervently held ideas by experiment. He expected that in a thin wire an electric current, considered to be inherently a conflict of opposing parts, would reveal a magnetic field. The theory was entirely wrong. But his demonstration of the actual production of magnetism by an electric current succeeded[14]—it was Oersted's passport to immortality. This synthesis set in motion the eventual elaboration of electromagnetic theory through Ampère, Faraday, and Maxwell, as well the invention of devices from telegraphy, motors, and generators to telephony and much else that is at the heart of modern industry. Thus, ironically, in tracing its ancestry, modern science and technology can discover a *Naturphilosoph* among their forbears.

Needless to say, Oersted's achievement had two very different, may I say conflicting, effects in the history of ideas. To the Nature Philosophers, it was proof, if proof they needed, of the correctness of their basic idea that nature was, contrary to Kepler, one coherent organism, a dynamic, pulsating playground of the basic force in its various guises, and infused by spirit. On the other hand, in physics it was soon reinterpreted and understood in terms of the field physics of Faraday, Maxwell, Helmholtz, Hertz, and others, all working in the Newtonian, even mechanistic tradition. Their triumphs helped put an end to the stranglehold

---

13. Ibid., chap. 19, "New Investigations into the Question: What Is Chemistry?"

14. Meyer, *Ørsted's Scientific Papers*, vol. 2, pp.214–218. See also pp. 223–245 and 351–398. There is a large amount of excellent scholarly secondary literature on this discovery.

which Nature Philosophy and the Romantic Rebellion had on much of the European imagination for decades. The great wave of Romantic science and philosophy submerged, at least for a time (with some parts of it emerging again in the fights around Darwinism later in the century).

From Spengler to Nazi Science

Throughout the twentieth century and since, there has been one sequel to Romantic science after another. This brings me to my second, far darker example of the periodic ascent of the Romantic Rebellion against well-established scientific ideas and methods. The new flight from reason and from the old order began to appear in such works as Oswald Spengler's apocalyptic book *The Decline of the West* (1918), in which soulless science was singled out as a cancer marking the inevitable, early end of our civilization, in preparation for the takeover of a new form of culture. In 1922, the great scholar of theology, Adolf von Harnack, spoke for many concerned intellectuals who saw and feared the trend: "Throughout the European world of culture of today there swells up again an international Romantic wave. . . . Instead of science and scholarship, one calls for 'Life' and for 'Intuition' instead of reason."[15]

Spengler's book, for example, had all the earmarks of the revolt: outrageous speculations, exciting predictions, distaste for all that the Enlightenment ideals represented.[16] Not surprisingly, most influential thinkers rejected it, but it found an enthusiastic audience in an initially obscure group of politically ambitious *Stürmer* who were intent on producing a new rupture in history, a new form of culture based on a refurbished *Sturm und Drang* ideology. They were of course the leading members of the National Socialist party in Germany. They sought out

---

15. Adolf von Harnack, *Erforschtes und Erlebtes* (Giessen: Alfred Töpelmann Verlag, 1923), p. 344.

16. Oswald Spengler, *Der Untergang des Abendlandes: Umrisse einer Morphologie der Weltgeschichte*, vol. I, *Gestalt und Wirklichkeit* (Vienna, Leipzig: Wilhelm Braunmüller, 1918); Spengler, *Der Untergang des Abendlandes: Umrisse einer Morphologie der Weltgeschichte* (Munich: C. H. Beck, 1980), which contains, in revised edition, both vol. I, *Gestalt und Wirklichkeit,* and vol. II, *Welthistorische Perspektiven* (originally published 1922); Spengler, *The Decline of the West*, vol. I (New York: A. A. Knopf, 1926) and vol. II (1928). Spengler's work and influence are discussed at greater length in chap. 5 of G. Holton, *Science and Anti-Science* (Cambridge: Harvard University Press, 1993).

Spengler personally, to recruit him to their cause. To his credit, Spengler rebuffed them.

But to succeed they didn't need him. In greater and greater measure, the populace at large opened their hearts to the message of these self-declared new leaders, a message that once again turned away from the core concepts of the Enlightenment. Those concepts, as Isaiah Berlin and others have pointed out, were unwittingly implicated in the rise of totalitarian tyrannies because the rebellions were engineered specifically against them. But the totalitarians, as so often, were also vastly helped by the ineffectiveness and tardiness of opposing forces to mobilize themselves. And while all the earlier, nineteenth-century rebels I have mentioned would have cried out in horror, this new group adopted some of the language and orientation of Romanticism—a fact contrary to the occasional preposterous allegation (e.g., by Zygmunt Bauman) that totalitarianism had its roots fully in modernity.

I thus will end tracing key episodes of the rise of postmodernisms by highlighting one of the darkest phases of the antimodern movement, insofar as it intersected with science.

The *Naturphilosophen,* no matter how misguided or confused some of their writings seem, had been as a group rather admirable opponents of the contemporary scientific worldview. Many were deeply learned scholars, or well-known poets, or serious scientists, and a few among the latter made contributions to *traditional* science despite themselves. But the more recent manifestation of the Romantic Rebellion to which I am now turning has been, in all these respects, the very opposite.

As Fritz Stern, Alan Beyerchen, Anne Harrington, and others have shown,[17] the National Socialist movement—even while holding on to some promodern elements, especially technology—was at its inception

---

17. Fritz Stern, *The Politics of Cultural Despair: A Study in the Rise of the Germanic Ideology* (Berkeley: University of California Press, 1961); Alan Beyerchen, *Scientists under Hitler: Politics in the Third Reich* (New Haven, CT: Yale University Press, 1977); Anne Harrington, *Reenchanted Science: Holism in German Culture from Wilhelm II to Hitler* (Princeton, NJ: Princeton University Press, 1996). See also Ute Deichmann, *Biologists under Hitler* (Cambridge: Harvard University Press, 1996); for a study of the place given to technology, see Paul R. Josephson, *Totalitarian Science and Technology* (Atlantic Highlands, NJ: Humanities Press, 1996).

largely rooted in various romantic ideas, whose common denominator was the rejection of much of modernity. Often these ideas were expressed as so-called *völkische* concepts, characterized by idealized notions of the nonrational fiber of the German people and a quasimythological premodern life style. These concepts, in part designed to provide a "meaningfulness" that modernism, for many, chronically lacked, were celebrated in mystical Germanic associations and Orders. The National Socialists built on that backward-looking, folk-oriented resentment of rationality. As Hitler explicitly stated in *Mein Kampf*, his ideal was the "*völkische Staat*." The scientist Philipp Lenard rejected much of modern physics in favor of the ether, which he associated with the seat of the German *Geist* or spirit. He and his fellow Nobel Prize winner, Johannes Stark, in common with many others, rejected the notion of scientific objectivity by claiming that the race of the researchers would determine their physics. For example, Stark's "German physics" would stress direct contact with nature, as against what he called "Jewish physics," which he charged with emphasis on theory and abstraction. A similar movement held sway in mathematics.

Out of the hellish welter of that movement I want to lift to visibility only one bizarre but telling and largely unknown example of this phase of the rebellion. In 1912, almost a decade before the first assertions of explicit Nazism in Germany, the Austrian engineer Hanns Hörbiger had his ideas published under the title *Hörbigers Glacial-Kosmogonie*.[18] Later known by the title *World-Ice Theory*, the book argued as follows: The world is under the influence of the eternal warfare between two conflicting principles, Plutonism and Neptunism. Correspondingly, there exist two types of celestial bodies with polar opposite character—hot ones, such as suns, and ice-covered ones. In the distant past, several of the latter type crashed into the earth, which, Hörbiger wrote, would explain a number of basic facts observable on earth now, as well as special historical events such as (of course) the destruction of Atlantis. Other ice bodies fell into the sun, with the resulting superheated water vapor explosively ejected; on cooling down, the vapor became cosmic ice, most of which formed the Milky Way, the rest falling to earth as hail.[19]

18. Published by Hermann Kaysers Verlag, Kaiserslautern.
19. For details of the theory and its influence, see Brigitte Nagel, *Die Welteislehre* (Stuttgart: Verlag für Geschichte der Naturwissenschaften und der Technik, 1991); and Robert Bowen, *Universal Ice* (London: Belhaven Press, 1993).

German scientists rejected this fable with scorn; but it became popular among general readers and eventually reached the highest echelons of the National Socialist regime. This point, I must interpolate, touches on my own main objection to the Romantic and antiscientific rebellion: namely, that—while it may subvert certain academic departments and media, divert students from a solid education, and for the masses act only as yet another opiate—*rebellion is most ominous, even deadly, when adopted by political leaders,* whether in Lysenko's Soviet Union, Mao's China, or here at home.

Among the National Socialist leadership, perhaps the most enthusiastic follower of the world-ice theory was Heinrich Himmler, graduate of Munich Technical University, Chief of the SS (*Schutzstaffel,* the security echelon), and proponent of the völkische, backward-looking aspects of National Socialism. Believing himself to be in contact with the spirit world, he wanted to replace Christianity with his own brand of a race-based, secular religion—a paganist revival of pre-Christian Germanic practices and beliefs with a strong admixture of ancestor worship. To further his purpose, he initiated a number of research institutes, foremost among them one called *Das Ahnenerbe,* dedicated to uncovering that heritage.

A whole department of that organization was devoted to the world-ice theory. In Himmler's mind, the theory connected with his notion that the so-called Aryans had descended not from early apes but from heaven, originating in sperms conserved in the cosmic ice that fell on earth. The ice mythology also suitably evoked the Nordic origin of the Aryans and the ancient Scandinavian epics. The resonance of these notions with Romantic thinking was also preserved, from Hörbiger's own writings on, in an inheritance from Nature Philosophy: the theory that the phenomena in the universe spring from a dichotomy of two basic forces—straight out of the Romantic preoccupation with polarity we have already seen in action before, but expressed in the world-ice theory in the dualism of fire and ice.

An equally sinister figure in German history, propaganda minister Josef Goebbels, had no reason to oppose these fantasies. His doctoral dissertation at the University of Heidelberg had the title "Wilhelm von Schütz: A Contribution to the History of the Drama of the Romantic School"— a title he later changed in his official biography to a more politically significant one, "The Spiritual-Political Movements in the Early Romantic

Period." Schütz, by the way, was among the least productive and most maudlin early-nineteenth-century Romantic poets. He was obsessed by the view that the innocence and piety that supposedly characterized premodern agrarian Germany was lost in modernity. His writings are replete with Romantic forests—the title of one of his plays (1808)—and with exotic voyages, occasional dionysiac frenzy, and a longing for the lost homeland. At any rate, it may well be that Goebbels, among the whole lot, was the one who most shamelessly and consciously manipulated the excesses of Romanticism for the advancement of Nazism.

The task of explicitly concocting a new spiritualization, to be spread widely by Goebbels and others, was given to Alfred Rosenberg, who later became foreign affairs secretary of the Nazi Party. His book, *The Myth of the Twentieth Century*, was the ideological bible defining the National-Socialist *Weltanschauung*. First published in 1930 and embraced by Hitler, who otherwise insisted he had no forerunner other than Richard Wagner, Rosenberg's vicious diatribe went through at least 130 editions and was meant to reach every household. It declared, "Our time, too, has its Romanticism," although one steeped in the glorification of force and *"Volksgeist,"* in racism, in what he called "the deep mystery of blood." Herder's late-eighteenth-century celebration of the people *(Volk)* and their national traditions *(Volkstum)* had undergone a grotesque perversion.

Much as one would like, one cannot avoid saying something about Hitler's own role with respect to science. By training and temperament he had no patience with the traditional Humboldtian style of education but rather abrogated it in favor of natural instinct—an echo of Herder's famous exclamation: "I am here not to think, but to be, feel, live!" Hitler wrote in *Mein Kampf* that the contemporary "semi-education severs the people from the instinct of nature."[20] According to Hermann Rauschning,[21]

---

20. From Hitler's *Mein Kampf* (Munich: Zentralverlag der NSDAP, 1939; first published 1925), p. 469. The passage, in translation, runs as follows: "Our semi-education severs the people from the instinct of nature, it pumps into them a kind of knowledge without being able to lead them to final understanding; industriousness and good will alone won't do; it has to be, necessarily, the understanding one is born with."

21. Hermann Rauschning, *Gespräche mit Hitler* (New York: Europe Verlag, 1940). Originally a follower of Hitler, Rauschning eventually became an opponent. Needless to say, his publication was later the subject of much debate. But whether or not Rauschning's reports all came directly from Hitler's lips, we shall see soon that the same sentiments pervaded the whole Nazi leadership.

president of the senate of Danzig, Hitler said, "I don't want there to be any intellectual education," and he proclaimed that mankind now found itself at the "end of the Age of Reason." Race was the carrier of natural instinct, which needed to be liberated from the dominance of reason. An admirer of Schopenhauer, Nietzsche, and the eugenicist Houston Stewart Chamberlain, Hitler bluntly rejected core scientific principles cherished by the German professors—objectivity, truth, respect for knowledge in its own right. He proclaimed, in sentences uncannily similar to what can be heard today from people quite innocent of the predecent:

> A new era of the magical explanation of the world is arising, an explanation based on Will rather than knowledge. There is no truth, in either the moral or the scientific sense. The concept of an independent *Wissenschaft*, free of any preconditions, could only emerge in the age of liberalism. It is absurd. Science is a social phenomenon. . . . With the slogan of objective science, the professoriat only wanted to free itself from the very necessary supervision by the State.
>
> That which is called the crisis of science is nothing more than the gentlemen are beginning to see on their own how they have gotten on to the wrong track with their objectivity and autonomy.

For Hitler, objective science was impossible, and all attempts in that direction should end. He claimed: "there can be only a science of a certain type of mankind and within a certain period. Thus, there is a Nordic science and a national socialistic one, in contrast to the liberalistic Jewish one."[22]

Not surprisingly, Hitler was also a great supporter of the world-ice theory; wholly in character, he planned to celebrate it in one of his grand architectural schemes, the transformation of the city of Linz into a new metropolis. In 1942 he discussed with Albert Speer its design, specifying a building that would contain "the three world pictures: Ptolemy's, Copernicus's, and the World-Ice Theory."[23]

Finally, Hitler's position with respect to science was to be made operational throughout the educational system by the notorious Bernhard

22. Ibid., pp. 210–211.
23. Ibid., p. 298.

Rust, Minister for Education of all Germany. Rust used the occasion of the 550th anniversary celebration of the University of Heidelberg in 1936 to explain, before an international audience, why the German authorities, from early 1933 on, had dismissed large numbers of scholars and scientists and changed the direction of the curricula fundamentally. As Rust put it in his talk, entitled "National Socialism and the Pursuit of Learning,"[24] you couldn't just change a few regulations. "It is our conviction that no significant reform in the pursuit of higher learning can occur except as it proceeds from a new idea of what science [*Wissenschaft*] really is. . . . It was necessary to act with all the more rigor and firmness, in that these individuals were seen to be using as a screen for furthering their own designs the prevailing theory regarding the pursuit of learning—namely, that it must be dispassionate, objective, free from prejudice and preconception." The "expulsion" of the scientists and the other scholars was necessary, because to act otherwise would, he said, only show "tolerance toward the arch-enemy of German self-confidence." In any case, many of those expelled were "of alien blood [who] were by nature incapable of conforming their teaching to the spirit of German culture." "Science . . . [is] not free, in that it is rooted in something other than science, namely philosophy." National Socialism's philosophical principles are the only basis on which science can find its "true objectivity."

Rust's lecture was followed by a speech by Ernst Krieck, soon to become rector of the University of Heidelberg. His lecture was entitled "The Objectivity of Science: A Crucial Problem." He dutifully repeated the main points of Rust, but took aim at Immanuel Kant in particular for "claiming for science complete autonomy as if it were a law unto itself. The whole struggle of tradition versus reconstruction centers around one crucial concept: the objectivity of science." One must go beyond Kant, Krieck remarked, because "an idea born of the Enlightenment—that is, an idea of Western civilization, bearing the marks of a limited period—has set itself up as an absolute and declared itself a criterion applicable to all peoples and at all times. Here we have an example of Western imperialism, a bold assertion of supremacy." "One can-

---

24. Translated from *Das nationalistische Deutschland und die Wissenschaft* (Hamburg: Schriften des Reichsinstituts für Geschichte des neuen Deutschlands, 1936). It contains the lectures by Rust and Krieck (discussed below) at Heidelberg.

not, like Kant, speak of 'mankind as such.' One must keep in mind the various fundamental racial characteristics of the people concerned."

Finally, Krieck ended with a warning to his unprotesting international audience: "We [in Germany] are called to lead the way. [It is] a path which our sister nations, some sooner, some later, are destined to tread." Starting three years later, a supine world discovered the cost of not having taken the new myth-makers seriously.[25]

Looking back over these very different historical examples of romantic rebellion—the periodic outbreak of the enthronement of the will and the rejection of reason and order—and finding similarities in our Age of Preposterism today, we must of course not conclude that past and present science wars are connected by a neat causal line. We must exempt the current Romantic Rebels from any suspicion that they have even heard of the predecessors I selected for discussion here or of their urge to see the end of science. Thus when a contemporary critic recently wrote that faith in the progressiveness of scientific rationalism has brought us to the point where "a more radical intellectual, moral, social, and political revolution [is called for] than the founders of modern

---

25. Because Krieck, now relatively unknown, was a type that defined the ruling elite of that era, a few more words about him illustrate (as do the speeches of Goebbels, Rosenberg, Rust, and others) the intellectual atmosphere governed by Hitler and his cohort. Ernst Krieck, originally a teacher in a primary school, was a fierce ideologist and prolific writer. A Nazi since the early 1920s, he was appointed to the chair of Pedagogy and Philosophy at the University of Heidelberg on April 1, 1934. His subsequent rise was irresistible. In mid-1935, upon the dismissal of the philosopher Ernst Hoffmann, Krieck became co-head of the Philosophical Seminar, together with Karl Jaspers. On September 30, 1937, he was made rector of the University of Heidelberg, a post he kept until October 1, 1938, when he "asked" to be relieved of these duties because his view on anthropology had annoyed Alfred Rosenberg. But Krieck remained in the chair of Pedagogy and Philosophy and wrote numerous books on National Socialist education. In *Wissenschaft, Weltanschauung, Hochschulreform* (Leipzig: Armanen-Verlag, 1934, pp. 6–7, 11), he wrote: "*Wissenschaft* does not create and does not live from a knowledge of truth that is independent of time and people *(Volk)* and valid in every time and every place. Rather, it brings a truth that is *völkisch* and temporally limited and enjoined by race, character, and fate, casting it into a rational form according to the worldview. . . . Like it or not: Nowhere shall we pursue the thing in itself, we shall no longer reach for the object of supposedly pure knowledge, and no longer merely record the state of the world. This fiction . . . is gone forever."

Western cultures could have imagined," she surely did not know that the same sentiments characterized the German ideologues. There are also marked idiosyncrasies in today's version, most notably the postmodern horror of Unity, unlike the penchant in favor of it in the nineteenth-century phase.

Despite all the differences between various romantic movements, no matter how benign or evil or even banal they may be in practice, in their very heart of hearts the Romantic Rebels of the past and of the present share that twofold sentiment I mentioned at the beginning: a pitiable longing for an idealized, spiritualized Golden Age assumed to have existed before modern science, and at the same time a more or less conscious determination to bring about a newly spiritualized future. That twofold sentiment, playing against the real and imagined failure of modernism to provide adequate answers to the psychological need for meaningfulness, is a chief source of energy and appeal of the movements—and probably always will be. Johann Wolfgang von Goethe hinted at this syndrome when he wrote, after studying the mechanical-materialistic book of Holbach, *System of Nature* (1770):

> But how hollow and empty we felt in this [book's] melancholic, atheistic half-night, in which the earth vanished with all its images, the heaven with all its stars. There was to be matter in motion for all eternity, and by this motion, right, left, and in every direction, without anything further, it was to produce the infinite phenomena of existence.... [And yet] we felt within us something that appeared like perfect freedom of will.

I started with Isaiah Berlin's chilling observation that no one predicted that our era would be dominated by the "enthronement of the will" and the "rejection of reason and order." I have sketched aspects of two earlier forms of the Romantic Rebellion to help respond to his grave question, "How did this begin?" But our historical excursion may teach us also that this variegated movement, if insufficiently attended to and its excesses only lazily opposed, may arise in ever-new guises, to assert dominance in the future, again and again.

— 13 —

Different Perceptions of "Good Science," and Their Effects on Careers of Women Scientists

The Romans had a special god of beginnings—Janus, with his two bearded heads facing opposite directions. I have always thought that the head looking to the left is scanning the past, while the other one is peering into the future. But both faces are usually depicted wearing a dark frown. I can understand why the left one seems unhappy: Maybe he is thinking of the mischief and tragedies that have marked so much of our past, and to some degree even persist in the present. But although I am writing on a sobering topic—this chapter might as well have the title "What You Thought Is Good Science May Be Dangerous to Your Career, Especially If You are a Woman Scientist"—it troubles me that the forward-looking face also frowns. In the end, however, I hope we will find the beginnings of a happy change on Janus's right face.

What Is Good Science, and Who Does It?

What is meant by good science? In the past, answers to this question were given with quick confidence. Immanuel Kant approved of two kinds of scientists, one looking for unities, the other for heterogeneity and diversity. Hans Christian Oersted also held that there are two kinds of good scientists, one seeking synthesis (discovering relations and analogies between phenomena), the other analysis (determining the disparity of things). In the later part of the nineteenth century, the naturalist Thomas H. Huxley had this stern admonition:

> Science seems to me to teach, in the highest and strongest manner, the great truth which is embodied in the Christian conception of entire

surrender to the will of God. Sit down before fact as a little child, be prepared to give up every preconceived notion, follow humbly wherever and to whatever abysses nature leads, or you shall learn nothing.[1]

Early in the twentieth century, the great chemist Wilhelm Ostwald distinguished between the classicists' and the romanticists' conception of very productive scientists, the former working slowly, and being a bit morbid in the bargain, the latter working quickly and with enthusiasm. And, finally, a more recent example: P. W. Bridgman's famous and, at least for me, clarifying observation that doing good science is just "doing your damnedest, no holds barred."

I am using these brief examples to hint that my own studies in the history of physical sciences soon made me aware of the great spectrum of answers to the question what "doing good science" may mean. This interest was reinforced by the personal experience of living and working among scientists and seeing their great variety of approaches. Nine Nobel Prizes have come to my colleagues in Harvard's Physics Department. Of course they all did superb scientific research; but in fact they were working in very different styles. That fact suggested a researchable question: Could one find a taxonomy of different styles of doing science?

And there was another puzzle, to which I became more and more sensitive. Until about two decades ago, all tenured professors in our department were male, and the department never had a tenured woman professor in the more distant past, either. (I hasten to add that at least it's different now—four of the recent tenure appointments were to women.) The relative scarcity of tenured women in physics throughout the United States—unlike in, say, biology—together with these puzzling observations motivated my decision, in the 1980s, to start a research program, called Project Access, to look seriously into the possible reasons for the gender differences in science careers.[2]

---

1. Quoted in Leonard Huxley, ed., *The Life and Letters of Thomas Henry Huxley* (London: Macmillan, 1900).

2. The results of this project were published in two books: G. Sonnert and G. Holton, *Who Succeeds in Science? The Gender Dimension* (New Brunswick, NJ: Rutgers University Press, 1995), and G. Sonnert, *Gender Differences in Science Careers: The Project Access Study* (New Brunswick, NJ: Rutgers University Press, 1995).

Defining Good Science by Studying Good Scientists

An intriguing part of our research, with Dr. Gerhard Sonnert closely acting as the Senior Research Associate, focused on a large group of promising scientists—about 700 men and women who had overcome earlier difficulties and had achieved a prestigious postdoctoral fellowship from a federal science agency. We asked what these scientists, initially about equally advantaged, think *is* "doing good science"? Second, controlling for variables such as age and field, do the men and women on average *differ* about what they mean by good science? Through the questionnaires and chiefly through our lengthy interviews with 200 men and women of the total group, as well as some comparative peer review of their work, we learned that on average there *are* differences between men and women in what they mean by good science. A third question: Can such differences in beliefs and resulting research strategies help explain the differences in their career outcomes?

Those differences in outcomes do of course exist; among the older scientists in this study, who ended up at top academic institutions, 88 percent of the men but only 43 percent of the women attained a full professorship. And in this elite group, the differences we found in their obtaining advances (in all fields except in biology) can be striking—on average there was, for example, as much as one full career step difference between assistant and associate professor, for the younger cohort in the physical sciences, mathematics, and engineering (PSME).

Our research, we thought, not only would have intellectual interest but could suggest policy changes in handling scientists, so as to ameliorate any unmerited differences in eventual career outcomes. And there beckoned also, once this phase of the research was finished, the possibility of going on to a study of what the typology of good science *now* may be between researchers, a typology much more sophisticated than those of Oersted, Huxley, and Ostwald. For example, and very much as a preliminary hypothesis, might there not be a whole series of types of notions of doing good science that characterize, either in the pure state or in some mixture, scientists today? For example: Type A, centering on the notion of challenging the prevailing scientific model or exemplar; Type B, centering on the hope to reach principle-oriented findings; Type C, positioning one's research in areas of basic scientific ignorance wherein advances might help solve social or national problems (what I call "Jeffersonian research"); Type D, emphasizing the applicability of already known science and engineering to technical and social problems; Type E, focusing on the synthesis of previously unconnected theories and findings; Type F, focusing on the potential for wide dissemination, recognition, and reward subsequent to the publication of scientific findings; Type G, rejecting "androcentric" or "Western" science and technology and seeking alternatives; and Type H, considering work in science and technology to be part of a social and political struggle for supremacy among rival research programs.

If some such typology could be found through actual empirical research and connected to the real-life outcomes, one could even imagine producing a test or a questionnaire by which—among other uses—a scientist on his or her own, privately, could determine to what blood type, so to speak, he or she belongs, and then look up what the career outcomes, on average, are likely to be. This last project is still only a plan. But let me now turn to what we have found in our research.

Using the questionnaires and interviews to study their career trajectories, we documented more clearly what we know intuitively: namely, that a successful scientific career is the golden result of an alchemical process involving dozens of ingredients—talent and merited opportunities, education, role models and mentorship, institutional backing, financial help or extra time given at just the right moment, persistence and ethical responsibility, and just plain good luck and serendipity. My task here is to deal with a narrow but essential component, how the individual's career outcome may be correlated with his or her particular perception of what is "good science."

One can well approach this question through the puzzling finding that the production of published papers by scientists in general is significantly larger for men than for women. This was true even in our highly select sample, where everyone had been certified as a promising scientist: The women scientists had an average of 2.3 publications per year, whereas men had 2.8, summing over all fields—in other words, the women published 20 percent less often than the men in our group. (In the biological sciences, the gap was smallest but still statistically significant.)

That general gender gap persists even if one looks at different types of publications. For example, in biology, our women on average produced about half a journal article a year less than men, and in physical sciences, mathematics, and engineering, the result was about the same. We were surprised to find, however, that women biologists published their work significantly more often than men in the form of book chapters, conference proceedings, and similar publications other than professional journals. One explanation might be that women biologists experience, or think they might experience, more difficulty getting research papers accepted in the ordinary peer-reviewed journals and so use other publication channels. But another, more intriguing possibility is that they choose these media because they may be more interested than men in publishing substantial, synthetic pieces that are too long for regular journals. This is a possibility we will consider further, shortly.

Some may ask: What difference does a smaller rate of publication make in terms of outcomes in academic rank? As we shall see, it is correlated with less advancement, and it is surely an even bigger factor in the careers of average scientists, those not as privileged as our group. A lesser rate of publication can be a vexatious handicap for any science career, since the length of the bibliography and a high rate of "production" are the easiest things to spot in a dossier, and for many people in position to decide on career advance, a candidate's long list of articles is the most impressive factor.

Historians of science, perhaps more than others, would be uncomfortable with any simple quantification of supposed quality. What matters most in the long run for the advance of a field of science itself is of course the *quality* of individual contributions, and that may not have a direct relation with mere *quantity*. In the extreme, one must not forget that Kurt Gödel turned all mathematics on its head with a total of about seventy-five printed pages. Yet when we sit down with a pile of dossiers of candidates for promotion or fellowships, aren't most of us prejudiced in favor of those who know how to churn out papers in abundance? Raw publication count, quite apart from its scientific quality, catches the attention of evaluators or future employers because it is simple, easily available, and widely used by others in the scientific community; for an evaluator outside the candidate's narrow specialty, there may be no other easy way to judge scientific productivity.

But if sheer counting of papers is too naive a measure, which nevertheless becomes important at decision points in careers, how are we to determine the *quality* of a given scientist? There is some helpful literature on the matter, for example, by Jonathan R. Cole and Stephen Cole, William R. Shadish, and others. Dr. Sonnert and I thought we might design a pilot project of our own to determine prevailing concepts of a scientist's quality—one we hope to expand someday, because we did get some results that surprised us.

We tried to replicate to some degree the real-life setting in which well-recognized senior scientists make initial evaluations of the scientific prowess of a candidate. We persuaded four senior biologists from two prestigious research universities to go over a set of twenty-four dossiers of male and female researchers in biology from our Project Access pool. We asked the panelists to assign quality ratings on the basis of what they had before them—the curriculum vitae, the bibliographies, and reprints of the six articles or chapters each individual thought represented his or her best work, either solo as a member of a team. After our evaluators had done their homework alone, we brought all four of them together to compare and discuss their judgments, just as they do so often in admission committees and in those over-heated chambers where tenure decisions are made.

Having learned from the dynamics we observed in the first round, we got six other distinguished faculty members of the biology department from two universities, three men and three women, to take on the task of evaluating the quality of the scientific work of forty-two men and women from the Project Access pool. We asked them to use a scale from 1 to the top grade of 5, similar to what many granting agencies do in assessing the merit of proposals.[3]

What was the final result? First, the average quality ratings given by our evaluators turned out to be significantly higher for the women biologists in our sample, an average grade of 3.67, versus 3.27 for men ($p = 0.0496$). This was not because the women among the evaluators favored women more than did the male evaluators. In fact, the women

---

3. The painstaking details of the pilot study have been described in an article by G. Sonnert, "What Makes a Good Scientist?: Determinants of Peer Evaluation among Biologists," *Social Studies in Science*, 25 (1995): pp. 35–55.

evaluators gave substantially better-quality ratings across the board, to both male and female candidates alike.

We then looked into the relation between the ratings given to individuals and all kinds of variables that might have predicted those quality ratings—for example, the prestige of the journals in which articles were published, or the selectivity of the graduate school of the individual, or his or her current academic rank. The biggest variable turned out to be, after all, publication productivity, with solo authorship adding to that predictor. On this point, our distinguished evaluators had reverted to type. Yet, while a rapid rate of publication, a least for the time being, still commands the biggest share of respect, the women candidates in this admittedly small study had emerged on average with higher-quality ratings. This difference persisted even after controlling for publication productivity and other variables found to influence quality ratings.

And there is another, interesting point. We did not tell our six evaluators that we had looked up, in the annual Institute for Scientific Information (ISI) Science Journal Citation Reports, the number of citations given to all the publications of the academic biologists in our sample. A tedious job; but we found that among the academic scientists, women's articles had received substantially more citations than the men's—that is, 24.4 versus 14.4 ($p = 0.0337$). Such a gender difference was also found by Scott Long,[4] in a larger sample in one subfield of biology. With all due caution about the use and abuse of citations in the literature, at the very least one can say that *in toto* the women's papers, on average, had greater visibility than the men's and, one may infer, were found to be of value to the field.

With their limitations, these data can tell us at least two things. One is that knowledge of this sort, if more widely understood, could well change how scientists judge and evaluate the quality of a candidate's scientific contributions. And the second point is that these findings place us at the threshold of another fascinating problem: What is it that women scientists, on average, *do* in their publications so that, even if not so numerous, those publications yield good and even superior quality ratings? To put it another way, is there, on the whole, a difference of

---

4. S. J. Long, "Measures of Sex Differences in Scientific Productivity," *Social Forces*, 71 (1992): 159–178.

some sort between what men and women scientists *think* is meant by doing "good science"?

On Styles of Doing Science

Let us look at some of the elements that enter into this business of trying to do good science. For example, *socialization differences* surely have some impact on the way science is done. Self-confidence, ambition, and related traits can favor any pursuit, not least in achieving success in science. But even in our elite group as a whole there was a substantial difference in that respect, with women reporting that they considered themselves as being "average" almost twice as often as did the men (35 percent versus 18 percent), and with men reporting themselves substantially more often "above average" (70 percent versus the women at 52 percent). About the same results emerged from the self-evaluation of technical skills, and even the way they thought others rated their scientific ability. More men than women describe their own overall scientific approach as "creative" (men 27 percent, women 12 percent)—an intriguing point, because it runs counter to the common stereotype of women as more creative and men as more logical and analytical.

Of course, all this may simply mean that the women on the whole were more realistic and sensible, their observations applying correctly to both sexes. Yet, hesitancy about one's ability, or at least the assertion of it, characterizes many groups who regard themselves as relatively new arrivals or, to use the sociologist Georg Simmel's concept, who think of themselves to some degree still as "strangers" in their milieu.

This possibility is made more likely by the finding that about four times as many women as men in our big sample said they had vague or unclear career aspirations when they started out in science. And in this observation, too, the women might have been simply more realistic, because they are quite likely to have encountered more career obstacles in their earlier days. Indeed, they might have a greater need for support when they encounter such obstacles; this would also follow from some theories that propose women have been socialized to depend on and to provide greater social connectedness—a characteristic that is not at the top of every department's list of grand achievements. In my researches in the history of science, I found that Fermi's magnificently productive laboratory in Rome in the 1930s was really run on a family model (see

Chapter 5), and one does more and more encounter a supportive atmosphere in American groups as well, particularly if they are blessed with success. But there is a long way to go before this model becomes common here.

Apart from whether gender differences exist in the way women work as scientists, our study found substantially more women than men thought that there *is* such a phenomenon. For instance, twice as many women as men (51 percent versus 26 percent) believed that their gender influenced their own professional conduct as scientists. More than three times as many women as men said their gender had an impact on the methods adopted in pursuit of their scientific projects.

These are of course only self-reports and are not necessarily related to outcome measures of scientific success, for men or women. And at any rate, these perceived gender differences were reported by only a minority of the men and women from the total population we interviewed. Yet for women who regard themselves as in some sense different from the bulk of the scientific professionals, such perceptions can have their costs, not least in terms of psychological stress and job satisfaction.

Now to our findings on women's professional conduct: in four words, women are less careerist. Among the men and women we interviewed, a very frequent observation in both cases was, not surprisingly, that male scientists, on the road to their career success, are more aggressive, combative, and self-promoting. We all know of terrifying examples of certain people with razor-sharp elbows. But it might well be that such conduct is also correlated with setting oneself more ambitious goals, in terms of scientific pursuit as well as career outcome. In surveying these findings, one can conclude that women scientists appear to be the "purer scientist"—they are somewhat less concerned with the political aspects of science, such as influence and power. Again, this connects with studies in the history of science, where the relative newcomer may be drawn to a field chiefly by its intellectual or aesthetic appeal. On average, our women respondents were more likely to see that scientific research is gorgeous, giving a lower priority to the hope to make a grand career by whatever means necessary. In society at large, women are at times under less social pressure in that direction, being less often the main breadwinner in the family. Of our group, women scientists were more likely than the men to emphasize the intellectual stimulation of science as one of the best things about scientific research. Ironically, when this joy is not matched with political savvy, it can slow one's career, at least in the short run.

One way to escape some of the rough-and-tumble competition of life at science's frontier is to choose a niche where, with luck, one can work on a problem that is important and yet not at the center of the volcanic eruptions of academic life. And that is just what our women had done more often than our men. It fits with the famous remark Marie Curie is said to have made when she was asked why she chose to work on what she later called radioactivity—a field from which even its chance discoverer, Henri Becquerel, had withdrawn: "I chose this field because there was no bibliography." This approach may also have a connection to our finding our women report that while in graduate school they were more collaborative, but after the postdoc position they became less collaborative than men, on average. Again, we know from studies of culture generally that those who previously had been marginalized often enter the mainstream through excelling first in a relatively noncompetitive portion of the field and find admission to a network of collaborators difficult.

In our study, we asked also about any epistemological differences in the way men and women scientists think. For example, do women refer to their own work as "non-androcentric" or "holistic vs. logical-sequential," the terms so often associated these days in the literature of some sociologists of science? In the responses from more than 100 women who were asked about this by our excellent interviewer, Sara Laschever, hardly any interest was detected in this matter. What did come through clearly was an entirely different response: that women scientists report to be more cautious and careful in their method and more attentive to details on the way to a conclusion. Numerous women acknowledged that they tended to be perfectionist in their work, not only to avoid failure and criticism, but more importantly to seek a broader, more comprehensive picture and produce more complete and synthetic papers. This relates, again, to the attempt to measure up to a higher ideal of what good science is, even at the cost of having to spend more time on one's project, instead of breaking up one's results into as many publishable salami slices as possible.

This finding is very suggestive, for it might help explain the publication productivity gap between the genders mentioned above. In the real world, the tendency—and it is only a tendency—for women scientists to linger over a problem to improve the quality of work at the cost of quantitative output is a noble trait, one well worth trying to teach men, too; but until everyone has learned that lesson, the pursuit of the noble may have its costs.

Now more on standards of good science. Many of the gender differences referred to so far (and especially the tendency of women on average to sacrifice quantity in favor of quality) led us to ask the respondents in our interviews directly what each thought characterized "good science." As noted, the men tended more to emphasize creativity. A good *presentation* of the findings was emphasized by men almost three times as often as by women. On the other hand, this ratio was entirely reversed when they spontaneously mentioned *integrity* as the *sine qua non* of the research process (men did so in 5 percent of the cases, women 14 percent). Finally, in line with the findings I previously noted, women volunteered *comprehensiveness* as a standard of good science almost twice as often as men. Women also, twice as often, emphasized comprehensiveness as a distinguishing mark between top-quality work and average work. In characterizing *bad* science, a striking difference was in the occasional mention of dishonesty, almost three times as often by women than by men (men 9 percent, women 24 percent). We must keep in mind, however, that in volunteered responses much that is taken for granted may not be verbalized.

In sum, we don't want to say that radical gender differences are emerging from these self-reports—that is, no neat, overall division between men and women was apparent from those long interviews and questionnaires. Indeed, there was plenty of evidence that a life in science is not all rosy for men either, and that many of our recommendations would apply also to better the careers of some of these men. But the main point of interest for us here was the reinforcement of the trends we have seen for women to be vocal about valuing broad or more comprehensive research projects that are also characterized by integrity and thoroughness, rather than preferring to gain a career advantage by higher output of what they might perceive as less serious work.

These self-reports and interviews also bolster our hypotheses for explaining the productivity gap, as well as the higher citation rate women's publications received on average in our pilot study. The picture that emerges is that the women scientists in our sample on the whole tended to uphold, to a statistically significant degree, more consciously what might be called the traditional standards of good science—the pursuit of fundamentals, with care, objectivity, and replicability. They tend to shun a more risky way of doing science, perhaps again in line with the general observation that relative newcomers to a field often feel they are being examined critically under a microscope.

Most of what I have presented here refers to findings of a project that lasted several years. Four final points need to be kept in mind. First, let me reemphasize that we chose to study the elite group in good part because if *their* members show gender differences, despite their having started with high promise on the same level, the differences are likely to be even greater among average scientists. Second, we concentrated chiefly on scientists in academe, because there the advancements and rewards are more uniform and easier to define, trace, and compare; in contrast, industrial or government-laboratory scientists have a great variety of differently marked paths, their publications may be internal reports, and so on. But we do not want to imply that leaving academic science is a negative move if it leads to a satisfying career doing science in other settings or serves science through administrative positions.

Third, the most successful women in our sample were, according to their self-reports and their careers, in most respects at the same level as the most successful men in the sample. I would hazard the prediction that, as time goes on, the same homogenization of rewards and risks is going to occur—is even occurring now, further down the chain, among the younger men and women in science. And I believe the younger women scientists are now also more savvy. In most science fields, the proportion of women is increasing steadily; some observers, such as Mildred Dresselhaus, have postulated that beyond a certain critical mass of women scientists—perhaps about 15 percent in PSME, and possibly more in other fields—the process of reaching parity among those of equal merit accelerates, as it already has done in most respects in biology, where the incoming groups are approximately equal with respect to gender.

And fourth, we have finally seen, in the last few years, ingenious interventions by policy makers, department heads, and the like—for example, making newcomers feel less isolated or providing visiting professorships and other programs to increase the visibility of talented researchers. Many demands from outside the science lab, such as family obligations, and relative isolation from the collegial network were reported even by successful women scientists. They are among the main disadvantages that characterized on average the career paths of our women scientists—the sorry result of various formal and informal obstacles leading to an accumulation of disadvantages,[5] both within academe and within the socialization patterns of the women themselves. Surely all these individuals want to be good to science. The question now is, how can science be good to them?

I end on a hopeful note. Over the past few decades, there have obviously been some steady improvements in the career expectations of women scientists as a whole. As Dr. Jewell Plummer Cobb has remarked, her advice to her mentees now is simply this: "Keep on going, things are getting better all the time." Science, in almost all fields, is becoming more and more fascinating in itself, and more fruitful for society. This, too, should better the lives of all scientists, male or female, if the supporting institutions do their share by cultivating that alchemy from which successful careers arise.

While much remains to be done, the darkest days are behind us. Looking again at that right head of Janus, as it gazes into the future, I think I see a trace of that long-delayed smile.

---

5. See H. Zuckerman, "Accumulation of Advantage and Disadvantage: The Theory and Its Intellectual Biography," in C. Mongardini and S. Tabboni, eds., *L'Opera di R. K. Merton e la Sociologia contemporeana* (Genoa: Edizioni Culturali Internationali Genova, 1989), pp. 153–176; and H. Zuckerman, J. R. Cole, and J. T. Bruer eds., *The Outer Circle: Women in the Scientific Community* (New York: Norton, 1991).

— 14 —

"Only Connect":
Bridging the Institutionalized Gaps between the Humanities and Sciences in Teaching

Academe today is characterized by two contrary movements. On one side we hear and experience the continuing strengthening of "interdisciplinary research." It is most notable in the natural sciences, where that research style flourishes as exciting new problems arise at the borderlines between established disciplines. Also, when basic research attempts to deal with societal problems—as illustrated in Chapter 11, under the convenient label "Jeffersonian Science"—it almost inevitably requires a coordinated attack involving several established disciplines.

On the other hand, much of teaching and administration in academe can still be described by the catch phrase "silence between the disciplines." Evidence may be found aplenty in scanning the course catalogues of universities, the way most appointments and promotions are handled, not to speak of the barriers within which most intellectual and social life is carried on among the faculty.

While hoping that the spirit of interdisciplinary cooperation will spread from the research laboratories and find a stronger place in the rest of academic life, I turn to the question how the perceptions and realities of silence between the disciplines came about, and what one may do about bridging the gap in the classroom.

Historically, the cultural dysfunction of disciplinary barriers is fairly new. Would the intelligentsia and the educators of earlier times not have been astounded by such a proposition? Would, say, Hermann von Helmholtz—physicist, biologist, physician, philosopher, and true *Kulturträger*—would he and his circle know what we are talking about? Or earlier, Alexander von Humboldt? Or earlier still, Madame Germaine de Stael, or Voltaire, or Émilie Du Châtelet (herself a mathematician, physicist, and philosopher)? Influential ideas on education, in their different ways, from Aristotle to Johann Pestalozzi, from Friedrich Schleiermacher to Alfred North Whitehead, from Francis Bacon to John Dewey and Philipp Frank, Robert Maynard Hutchins, James Bryant Conant, and others, were designed to prevent that interdisciplinary silence. So is the silence that still exists perhaps a by-product of more recent historical, cultural, and social tendencies?

Specialization as Twentieth-Century Worldview

Some saw the trend toward academic specialization coming early. The American historian Henry Adams, writing his autobiography in 1905, warned that the course of history in the new twentieth century would move away from the vestiges of cultural unity—a unity which he assumed to have been based on the centuries-long hold of religion in the West—and tend toward fragmentation and disunity, multiplicity, a state symbolized for him by the violent forces in the newly discovered radioactive elements.

The intellectual and emotional cost of modern professional specialization was touchingly described by Lionel Trilling in his book, *Mind in the Modern World* (1973, pp. 13–14):

> Physical science in our day lies beyond the intellectual grasp of most men. . . . This exclusion of most of us from the mode of thought which is habitually said to be the characteristic achievement of the modern age is bound to be experienced as a wound given to our intellectual self-esteem. About this humiliation we all agree to be silent; but can we doubt that it has its consequences, that it introduces into the life of the mind a significant element of dubiety and alienation which must be taken into account in any estimate that is made of the present fortunes of mind?

The scholar of modern history Fritz Stern, deploring that he could not grasp the "genius" behind Einstein's work in physics, voiced his own dismay on this point: "I felt this exclusion the more as I came to realize the intensity of the aesthetic joy that Einstein and his colleagues found in their discoveries, as their correspondence exemplifies. We are shut out from that knowledge and from that particular beauty. Lionel

Trilling was abundantly right in calling this exclusion an unacknowledged wound."[1]

In a less elegant but more concise way, the narrowing of attention on the highly specialized but fascinating task at hand was encapsulated in a remark by molecular geneticist Eckart Wimmer (*New York Times*, July 19, 2002): "Every minute you don't work, you lose out on science."

Perhaps the most profound attempt to define the source and tragic costs of a dysfunction among the specialties was made many decades ago. In May 1935, the distinguished German philosopher Edmund Husserl, no longer allowed to lecture and publish in his own country, was invited to speak at the *Kulturbund* in Vienna, and later in Prague. His lecture was subsequently expanded in a book with the title *Die Krisis der europäischen Wissenschaften und die phaenomenologische Philosophie* (published posthumously in 1954). Much of his argument has relevance for our topic, not only for education but also for the current debate about the emergence of elements of a national identity. Indeed, the title of Part I of Husserl's book is nothing less than this: "The Crisis of the Sciences [*Wissenschaften*] as Expression of the Radical Life-Crisis of European Humanity."

Husserl was painfully aware in the 1930s of the "profound malaise among the educated," "a deeply felt lack of direction for man's existence as a whole—a feeling of crisis and breakdown," and not merely because of the then-prevailing political conditions, not merely because of what he called the "irrationalism" among the educated. In a letter at that time he warned of the "complete upset [in] the international community [of] a harmonious unity of the life of nations, with its source in the rational spirit." To Husserl, the schism between the *Wissenschaften*, the lack of a widely shared culture, seemed to doom the rise of what he called "the primal phenomenon of spiritual Europe"—a vision of Europe in which Husserl significantly included the United Kingdom and the United States of America.

What had come to full fruition, in this view, was the long-range damage unwittingly introduced by Galileo's method of science. For that method had begun to inject, as already prophesized by John Donne in

---

1. F. Stern, "Einstein's Germany," in *Albert Einstein: Historical and Cultural Perspectives*, ed. G. Holton and Y. Elkana (Mineola, NY: Dover Publications, 1997 [1982]), p. 320.

1611, the schism in what Husserl called a coherent *Weltanschauungsphilosophie*. Not only did this schism manifest itself in the centrifugal forces that tore, Husserl said, "the total worldview of modern man" into "split disciplines." The separate specializations even caused the methods used in those fragments to ape the positive sciences that seemed "so unimpeachable within the legitimacy of their methodological accomplishments." Thereby the humanities had become powerless to deal with the most fundamental questions of all, *"the questions which are decisive for a genuine humankind."*

One may quarrel today with some aspects of his analysis, but in my opinion the net result is correct to this day: Settling for the "split disciplines" would impede the ability and duty of intellectuals and other educated persons to attend properly to their main task—which is to contribute, in the various necessary ways, to the health of "a genuine humankind."

"General Education" as a Twentieth-Century Remedy

If we now set out, in the face of Husserl's pessimism, to build at least into the *educational* process some of those much-desired bridges between the disciplines, certain tested models of the past come to mind. How about re-instituting, in liberal arts education, the trivium and quadrivium, the courses of study used for centuries, ranging from logic to astronomy and music? One might call it the *encyclopedic solution*, meant to assure that future Henry Adamses and Eckhart Wimmers will find common ground and harmonious converse.

Variations of this encyclopedic solution persisted into the nineteenth century. For example, in a letter of September 7, 1814, Thomas Jefferson, during his retirement at Monticello, wrote to Peter Carr, who had evidently inquired of just such an educational program. Jefferson obliged, giving a detailed list of subjects for the instruction of those who "aspire to share in conducting the affairs of the nation"—the persons Plato had called the magistrates. Here, much abbreviated, is Jefferson's list:[2]

---

2. Pp. 1349–50 in Jefferson's *Writings*, The Library of America, 1984. (For related lists, see pp. 462–463 and 1422–25.)

Languages (including of course Latin and Greek); history, ancient and modern; *belles lettres,* even with special help to be given to the "deaf, dumb, and blind."

Mathematics, including fluxions or calculus; physics; chemistry; mineralogy; biology; zoology; anatomy; the theory of medicine; and meteorology.

Philosophy, including ethics, the law of nature and nations, government, political economy.

This list was intended as just a start. For those who were to graduate on to a professional school, they would have to add such subjects as architecture, gardening, painting, sculpture, law, theology, and much more.

Why so many subjects to study? Because, first and last, Jefferson believed, as did many after him, that only a people widely educated in a spirit of free inquiry could govern itself and flourish in a democracy. Today, one would add that such an education prepares one to deal better with complex social problems.

The same Jeffersonian spirit, though less exhaustive, animated later educational experiments, such as those in Contemporary Civilization at Columbia University and the Great Books programs at Chicago and St. John's College.

Leaving aside other efforts to find common ground among the disciplines—for example, by the Vienna Circle under Moritz Schlick, its Berlin counterpart under Hans Reichenbach, and the *International Encyclopedia of Unified Science* of Otto Neurath—I turn to one in which I participated some decades ago, with all the enthusiasm of a young faculty member. The idea for it came from the top. Harvard's president at the time was James Bryant Conant. While holding prominent positions in Washington during World War II, he had been startled by finding that young soldiers with little knowledge about the values characterizing a free society had been sent into bloody battle to preserve it.

At his instigation, a Harvard faculty committee published a plan in 1945, significantly called "General Education in a Free Society" (the famous Red Book), a manifesto meant for all secondary schools as well as universities. (That makes this example less parochial.) At Harvard and many other places, General Education initiated, for the first two years of college, a series of courses, each a one-year course intended to present a serious overview of the main achievements in knowledge and sensitivity—at least one in the natural sciences, one in the social sciences, and one in the humanities. The subtext of the whole program was chiefly to turn out a citizenry that would have the rudiments of education in each major field; that would be aware of and treasure the heritage of our civilization; and that would, by providing "a common learning for all Americans as a foundation of national unity," meet the "supreme need of an America education," namely, "a unifying purpose and idea."

Indeed, while it lasted, the whole program could boast of many grand courses that attempted to lay out the unities behind each major segment of knowledge and provide a "shared experience" for the student body. The faculty, too, faced with this new and ambitious task, had to subject itself to a serious expansion of its initial interests. And to signal to the faculty his seriousness, Mr. Conant himself, while president, undertook to teach one of those General Education courses.

From the Bridge to the Cafeteria

But after some years, the initial enthusiasm waned. Even a program led by distinguished intellectuals such as Conant, Erik Erikson, David Riesman, Paul Tillich, and George Wald succumbs eventually to what Max Weber called the routinization of charisma. The spirit of fragmentation reasserted itself in the faculty, driven chiefly by the professional embrace of sub-specialization. A rebellion against General Education at Harvard and elsewhere, led first by some scientists, urged a return to concentration on current knowledge in specific specialty fields, forsaking the broader, historical, and humanistic approach initially launched by Conant. In time, the revised program developed into ever-narrower courses, on the way to a new mandate called the Core Curriculum—one also widely copied throughout America (although, after a reign of some thirty years, it is being reconsidered here and there). As one of my colleagues remarked at the time, it was a "departmental takeover" of the General Education program.

When the Core Curriculum was announced, on the very first page of our College's Catalogue of Courses, its proud aim was to offer courses that "do not define intellectual breadths, as in the mastery of a set of great books . . . [but] rather seek to introduce students to the major *ap-*

proaches to knowledge [italics in original] in different specialties": in short, the goal was not primarily the pursuit of knowledge but the demonstration of different ways of thinking.

In practice, this high-sounding purpose presents some problems for a would-be builder of bridges across the separated elements of our culture. For example, under the heading "Literature and Arts," our students are asked, depending on their concentration, to select as few as one out of nearly sixty different, one-semester (thirteen weeks) courses. These have titles ranging from "Fairy Tales" to "Dante," from "American Jazz" to "Recollecting the Way of Life in Pompeii." For many students, a single course from this list would be their only required contact with literature and the arts during their four years of college. Individually, some of these courses may be very well done; but the program as a whole hardly allows for enough depth, scope, and time to nourish a lifelong interest.

As to the natural sciences, for which the new Core Course legislation specifically outlawed any historical components, there is a similar large spectrum of mainly narrow one-semester offerings, from which every student not heading for a science career must select two. The result is that a little less than 6 percent of our non-science students' total educational experience in college is reserved for science and technology. To be sure, this paltry fraction is still better than at most other colleges in the United States today, which have largely dropped all structured systems and let each student assemble his or her own program. Indeed, at present, only 30 percent of all colleges in the United States require even a single hour of science for graduation—building on a largely dismal secondary school education, about which the less said here the better.

In short, in U.S. college education there has been a visible abandonment of the quasi-encyclopedic approach of the mid-twentieth century. It is not too much to say that the metaphor of the Bridge has been replaced by that of the College Cafeteria. This model fits with the postmodern current, still strong in American academe, which is directed against any canon and assigns equal authority to each fragment of interest, be it the epics of the Trojan War or, at one well-known college, a course, in its Music Department, on "Humming."

The result, therefore, in most American higher education today, is the *institutionalization* of the silence between the disciplines, beginning at the college level. Rarely is the faculty challenged to go beyond its main professional preoccupation; and rarely do two students share some, if any, of the intellectual experience of their four years of college. Gone are the four characteristics that, according to Daniel Bell's analysis,[3] were common to the General Education programs of half a century ago, as well as earlier programs at Columbia University and Chicago and their many imitators:

1. Some attempt at consensus, "instilling in students a sense of common tastes, though not necessarily a single purpose."
2. Some attempt at awareness of tradition, of the history of Western civilization, "its moral and political problems, the travails of the idea of freedom," and with it instilling some "idea of civility."
3. Going beyond specialization: In addition to offering thorough pre-professional training, establishing a parallel but wider track to broaden the specialists' view to encompass "humanitas."
4. Integration: As a balance to the "staggering expansion of knowledge produced by specialization," an emphasis in those General Education and similar courses on "the broad relationship of knowledge, rather than [focusing only] on a single discipline," based on "the underlying assumption . . . of the need of an *interdisciplinary* approach."

I know of only a mere handful of college programs in the United States today that still try to follow those guiding principles.

New Approaches for the Twenty-first Century

At this point you may wonder whether I can offer any positive response to the problems I have sketched. I say yes. Even now, some versions of the earlier, honorable models I have cited may flourish in proper hands. There are also other pedagogical approaches, both for secondary schools and colleges or universities, that might ameliorate the institutionalized silence between disciplines. I shall describe briefly three personal experiments in designing such courses, necessarily centered on physics—although (and this is important here) the same principle is worth experimenting with in any field.

---

3. In his book, *The Reforming of General Education* (1966), pp. 51–52.

1. As in all my examples, one has of course to insist on conveying a sound knowledge of the main subject matter. But the easiest and least demanding policy to expand on the basics is to diverge at the right moment from teaching—for example—pure physics, and *add key elements of some of the neighboring natural sciences*—astronomy, astrophysics, chemistry, biophysics, and certain aspects of technology (ranging from transistors and radar to magnetic resonance imaging, with acknowledgment of the technological advances in early China and Islamic nations). Such an extension, demonstrating the lack of silence between at least segments in the natural sciences, is by no means artificial. For example, a Physics Department such as Harvard University's has, on its senior faculty, a large spectrum of scientists with diverse but overlapping interests—including those who have joint appointments with other departments, in molecular biology, chemistry, electrical engineering, astronomy, mathematics, chemical biology, and the history of science. In short: at the level of *research*, interdisciplinarity is now the key to some of the best new achievements.

2. I turn to a second type of course that aims to prepare students for an appreciation of the other aspects of our culture while at the same time teaching, say, good science, and show it to be the result of a more general on-going human adventure. This second type of course involves adding to the first exemplar *an ordering of the sequence of successive science topics of the course in the historical sequence of its actual developments*. It permits tracing the evolution of a science through painfully achieved advances; through frequent struggles with errors, conjectures and refutations; through the complex interplay between theory and experiment. Moreover, it inevitably brings in some philosophical, social and other cultural factors that helped (or interfered with) scientific advances.

When inserted at particular points in the course, this approach, too, does not add greatly to the time burden. But it does add an element that for many students is humanly interesting, alerts them to the "cultural soil" that nourishes science, and is also helpful in the students' own complex struggles with initially counter-intuitive scientific concepts. Let them see how some of those giants experienced troubles similar to their own. For example, Isaac Newton early on poorly understood his concept of inertia to be a *vis insita*, an inseminate force in the moving object—as if a little person inside the body kept pushing it to keep it moving. (So do many students initially, today.)

Introducing the history of science into science courses is fairly new in pedagogy, although there exists a small group that has published texts in this spirit.[4] But at least in the U.S.A., this enlarged approach has been strongly supported by our National Academy of Sciences, in a major report (of 1996) on proper science teaching standards, first of all at the secondary (high school) level. I quote from the booklet *National Science Education Standards* (p. 107): "In learning science, students need to understand that science reflects its history and is an on-going, changing enterprise. The standards for the history and nature of science recommend the use of history of science in . . . science programs, to clarify different aspects of scientific inquiry, the human aspects of science, and the role science has played in the development of various cultures." The Academy's volume then shows in detail how this might be done.

Another significant benefit derived from including the historical aspects of science in science courses is that it demonstrates by example *commonalities* between persons engaged in scientific work and creative people on the other side of that famous gap—poets, composers, artists, and others. The commonalities I refer to are the frequent use, during the nascent stages of scientific work, of metaphors, analogies, and themata, of the visual imagination, and of occasional daring leaps of the imagination unsupported by prior experience or current consensus.

3. My last example of a course that can bridge the gap between disciplines exploits cross-cultural potential even more seriously—and therefore is to me the most interesting. This type of discipline-centered course from time to time devotes a lecture period, after appropriate reading assignments, *to interactions among the sciences and relates scientific advances to social and other cultural trends and products*. Examples can be readily found in the first and second industrial revolutions, based respectively on steam and electricity; in the effects by and on physical science of the electronic deluge; in the occasional, fascinating interac-

---

4. For examples of this mode of teaching, see G. Holton, *Introduction to Concepts and Theories in Physical Science* (1952 and later editions), G. Holton and S. G. Brush, *Physics, the Human Adventure: From Copernicus to Einstein and Beyond* (2001), and D. Cassidy et al., *Understanding Physics* (2002).

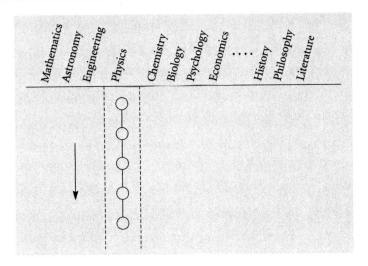

Figure 8. Traditional presentation of topics in introductory physics.

tions between science and literature as well as philosophy; and in the ethical problems scientists may face in their work.[5]

That approach to teaching ameliorates a grave difficulty with most other programs. The difficulty is this: In practice, as a student sees it, even the most inspiring set of general survey courses, each covering a major field, offers essentially a two-dimensional array with the chasm between disciplines built in from the start. Each of these courses presents its topics as a string of pearls arranged in some logical order (Fig. 8). In the total program consisting of several courses, these separate strings hang side by side, touching one another barely, if at all. Next to the string of topics covered in a science course, there may be—in another building—a course on literature, with its own cascade of linked jewels, each jewel standing for a week's presentation; and next to that hangs another filament bearing the successive kernels of social science; and so on. But while it may have served its purpose some time ago, that two-dimensional model of teaching no longer corresponds to the reality

---

5. An extensive discussion of the educational philosophy and practice for such courses is given in chap. 15, "Physics and Culture," in G. Holton, *Thematic Origins of Scientific Thought* (Cambridge: Harvard University Press, 1973 ed.).

and necessity of the world today. The total scope of culture today is not two-dimensional but multi-dimensional, a kind of web or patchwork of overlapping and diverse elements, some complementary to one another, some even contradictory.

A Network of Knowledge

To illustrate this point, let me conjure up the idea of a dinner party among friends—for, as J. Robert Oppenheimer once wrote,[6] while we can't share all knowledge, "We can have each other to dinner. We ourselves, and with each other by our converse, can create, not an architecture of global scope, but an immense, intricate network of intimacy, illumination, and understanding."

At this imaginary party, one might seat a physicist who is also deeply interested in philosophy and a passionate reader of military history next to a sociologist who happens to be also a collector of Japanese prints. Next to him is a cellist whose undergraduate degree was in biochemistry; and her neighbor is a judge on the Judicial Court who is always eager to hear the latest news in science, not only because a larger and larger portion of the caseload before him involves scientific and technological matters.

What binds the guests together is not that they fully share one common culture, or that each is ready to recite a sonnet of Shakespeare after stating the Second Law of Thermodynamics. However, within groups of alert people with wide-ranging interests there are *sufficient overlaps* of significant elements of knowledge and expertise, of taste, above all of open curiosity. Together, they form a multi-dimensional array.

To make this example more visual, imagine a crossword puzzle in which the letters in the words on each of the horizontal lines stands for the various active interests of one person. But each horizontal line of letters intersects, orthogonal to it, with one or more other vertical lines. Together, these vertical lines also represent the several active, life-long interests of other educated persons. It is this overlapping at intersections of lines, this intercalation of a variety of different elements—rather than one common unity—which is at the heart of our modern culture. Al-

---

6. In *Daedalus*, Winter 1958, p. 76.

though this network is a fragile compromise, it keeps a culture from decaying into a Tower of Babel. Nobody escapes his or her individual limitations, but everyone in such a group can know and feel enough to be a member of an intellectual network. Indeed, perhaps deep down this has always been so.

The imaginary dinner party I just described is, as you may have guessed, a metaphor for what I hope modern pedagogy, and especially the third type of education, can do to encourage the production of more communities of different specialists with greatly enlarged horizons, whether in academe or other professions, in industry, in governance, and so forth.

To approach this goal, the traditional way in which we have taught our disciplinary courses is patently insufficient. The traditional course values chiefly training, not orientation, whereas historically, most basic findings have developed not linearly but as part of a constellation of an interdisciplinary network. What I am recommending, especially in the third type of course, is what I call a *connective approach* to the teaching of science, and indeed to the teaching of each field, not least as intellectual preparation for the student's later life.

Attending to the Mosaic of Culture

I end with an example for classroom use, in this spirit, of just one topic for a week or two: the Newtonian synthesis, based on the works of Kepler, Galileo, Descartes, and their contemporaries (Fig. 9). We can show that Newton's ideas were much influenced by debates on the nature of physical knowledge going back in time to the ideas of the Greeks (line A). Conversely, the success of seventeenth-century physics had a striking impact on later philosophy *(F)*, on Immanuel Kant, on the conceptions of the separation of primary and secondary qualities, and on the mathematization of reality that haunts parts of sociology and philosophy to this day.

Again, among the giants on whose shoulders Newton knew he was standing were Greek mathematicians *(C)* such as Euclid and Apollonius. In turn, Newton's mathematics made that field flourish later *(D)*. The philosophers of the Age of Reason were of course also deeply influenced by Newton's physics *(F)*, as was, in his way, the founder of a

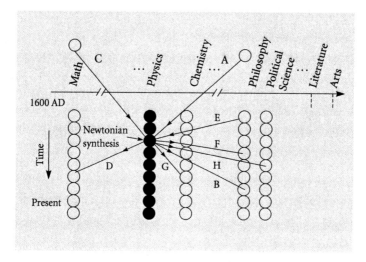

Figure 9. Connective presentation of a topic in physics.

branch of chemistry *(G)*, John Dalton. Turning to political science, one can refer to the explicit acknowledgment of the debt to Newtonian science in the "balance of power" imagery that was used in drawing up the Constitution in revolutionary America *(H)*. Other connections—such as to literature, when Newton swayed the Muses, and to art—can be referred to easily and bolstered by reading assignments. The course should make it clear that those links between science and other fields were not unilateral, but worked the other way, too. For example, Einstein's formulation of the relativity theory can be shown to have benefited from his reading in Hume, Kant, Ernst Mach, Spinoza, and Goethe.

There are many other potential examples that might be used in such a course, which should be based on the presentation of both sound discipline and connective links to the rest of our cultural tradition. In the end, the course must present not only that string of pearls, all of them within one field, but at the very least a good glimpse of a tapestry of cross-connections among many fields. Moreover, it can prepare students for the possible need to do interdisciplinary work in real life.

One might now be thinking that only a new Erasmus could run such a course. That is not correct, for two reasons. First, each instructor interested in this approach need not assume his or her version of the course—and they can all differ—has to be fully developed the first time it is offered. The instructor can build up segments, year by year—as many whom I know have done. Second, just as in a typical research laboratory, much of the work, with some guidance from the top, is done by students, so here too, instructors do not act alone. They give the class just enough background to legitimize the approach and interest the students, and then assign them papers that will lead to further exploration on their own. To this end, one aids the students by supplying lists of references and other materials, prepared for this purpose.

To be sure—unless they are helped by subsidized tutorials—it takes courage for instructors to prepare themselves to convey an antidote against silences between the disciplines. It would be my hope that the academic institutions might institute programs to help selected, interested teachers or faculties to experiment with developing such courses, for example, by providing resource publications and teacher training.

What I have proposed, and I have tried in class and teacher training, is not going to repair quickly what Husserl bemoaned as the loss of that supposed "harmonious unity of the life of nations." But students taught in such courses may emerge attuned to the more modern, more practical view of culture as an assembly of different portions of the whole—a mosaic rather than the icon of a great, commanding edifice such as the cathedral at Chartres, which haunted the imagination of Henry Adams.

The time has long come to repair our comfortable, old-style curriculum and attack its inherent, institutionalized silence between the disciplines. For behind the mere question what to talk about in our next set of class lectures, there loom, since antiquity, and never more than now, "the questions which are decisive for a genuine humankind."

Acknowledgments

Index

Acknowledgments

Chapters in this book were based on earlier publications or presentations, as follows:

Chapter 1, *Daedalus* (Fall 2003), vol. 132, no. 4.

Chapter 2, Lecture at inauguration of Einstein exhibit at American Museum of Natural History, New York (May 2003).

Chapter 3, *Physics Today,* vol. 53, no. 7 (July 2000). Reprinted with permission.

Chapter 4, *Los Angeles Times Book Review* (October 17, 1999), and essay in M. Doerries, ed., *Michael Frayn, Kopenhagen* (Göttingen: Wallstein Verlag, 2001).

Chapter 5, *Times Literary Supplement,* 5154 (January 11, 2002).

Chapter 6, *Behaviorology,* vol. 5, no. 1 (2000).

Chapter 7, *Physics Today,* vol. 52, no. 9 (September 1999). Reprinted with permission.

Chapter 8, The Paul Tillich Lecture, presented at Harvard University (April 12, 2004).

Chapter 9, *Leonardo,* vol. 34, no. 2 (2001).

Chapter 10, Article on "Themata" in *Dictionnaire d'Histoire et Philosophie des Sciences* (Paris: Presses Universitaires de France, 1999).

Chapter 11, L. Branscomb, G. Holton, G. Sonnert, eds., *Science for Society: Cutting-Edge Basic Research in the Service of Public Objective* (Cambridge, MA: Kennedy School of Government, Harvard University, May 2001).

Chapter 12, *Journal for History of Ideas,* vol. 61, no. 2 (July 2000).

Chapter 13, Chapter in C. Selby, ed., *Women in Science and Engineering: Choices for Success,* Annals of New York Academy of Sciences, vol. 869 (New York: New York Academy of Sciences, 1999).

Chapter 14, Lecture at Conference on "Silence between the Disciplines," Berlin (2002).

Index

1919 eclipse, 27
1/*v* law, 59

Abiko, Seiya, 18n1
Acceleration, 6, 107
"Accumulation of Advantage and Disadvantage: The Theory and Its Intellectual Biography" (Zuckerman), 193n5
Across the Frontiers (Heisenberg), 8, 33n7
Acta Mathematica, 115n1
Adams, Henry, 197, 208
Adams, James Luther, 97, 114
Advancement of Science, and Its Burdens, The (Holton), 151, 156n6
Age of Preposterism, 163
Age of Reason, 177, 206–207
Ahlgren, A., 148
AIDS, 159
Aiken, Howard, 70
Akademie Olympia (discussion circle), 12
Albert Einstein: Historical and Cultural Perspectives (Holton and Elkana), 196n1
Albert Einstein: Philosopher-Scientist (Schilpp), 31n6, 119n6, 150n8
A l'infinitif (Duchamp), 125
Alpha particle, 56
Alpine University, 104
Amaldi, Edoardo, 53, 55, 57–58, 62–63
Amaldi, Ginestra, 59
Amaldi, Ugo, 53
American Association for the Advancement of Science, 74, 148
American Historical Review, 42n11

American Journal of Physics, 22n4, 83n3
American Philosophical Society, 83
American Physical Society, 83
Ampère, André Marie, 171
Analysis of Sensations (Mach), 67
Anarchism, 118
Anderson, Carl, 55
Anderson, Herbert, 62
Anderson, Perry, 163n6
Annalen der Physik, 22
Antheil, George, 124
Anti-Galileans, 91
Anti-Semitism, 21, 28, 32–34, 108, 174
"Apotheosis of the Romantic Will, The" (Berlin), 162
Aristotle, 98, 144, 149, 195
Art, 207; Duchamp and, 117–118, 125–130; *n*-dimensionality and, 119, 124–131; positivism and, 132–133
Artificial radioactivity, 54–62
Astronomia Nova (Kepler), 164
Atheism, 9–10, 108
Atlantic Monthly, 90
Atomic Histories (Peierls), 40n6, 42n11
Atomic science: atomic bombs and, 33–35, 39–42, 90; Copenhagen legend and, 36–47; *Uranverein* and, 38, 40–42; Jungk and, 38–44; morality issues and, 39–41; Powers and, 41–42; Fermi and, 52, 54; artificial radioactivity and, 54–62; beta decay and, 55; neutrino hypothesis and, 55; proton-neutron model and, 55; positrons and, 56; H-bomb and, 90; Copenhagen interpretation and, 106

Atomism, 142, 150
Atoms for Peace conferences, 90
Atoms in the Family (Fermi), 58n1, 64
"Autobiographical Notes" (Einstein), 3–4, 6–7, 18n1, 21, 23, 109
"Autobiographical Reflections" (Tillich), 100–101

Bacon, Francis, 143, 154–156, 195
Ballets Russes, 133
Banquet Years, 133
Barkla, Charles Grover, 50
Barth, Karl, 104
Becquerel, Henri, 190
Begegnungen (Tillich), 114
Behaviorism and Logical Positivism (Smith), 68n4
Behavior of Organisms, The (Skinner), 70
Bell, Daniel, 201
Benchmarks for Science Literacy, 148
Ben-David, Joseph, 10
Bergson, Henri, 60, 134
Berkeley, 56
Berlin, Isaiah, 2, 4, 144, 162–164, 173
Bernardini, Carlo, 63
Bernardini, Gilberto, 55
Bernstein, Jeremy, 39n4, 40n6, 41n7, 46, 84n5
Berzelius, Jöns Jacob, 170
Beta decay, 55
Bethe, Hans, 53, 55, 64
Bewusstseinswandel (von Weizsäcker), 40n5
Beyerchen, Alan, 173
Beyond Freedom and Dignity (Skinner), 71
B. F. Skinner and Behaviorism in American Culture (Smith and Woodward), 66n1
Bible, 11; Einstein and, 3–4; Rabi and, 84–85
"Biblical Religion and the Search for Ultimate Reality" (Tillich), 96, 114
Biographia Literaria (Coleridge), 37
Biographical Memoirs (Fellows of the Royal Society), 22
Biographie (Frisch), 36
Biologists under Hitler (Deichmann), 173n17
"Biology and Social Behavior" (Wald), 69

Birkhoff, G. D., 74
Bitter, Francis, 86
BKS theory, 28
Blake, William, 166
Bloch, Felix, 55
"Boats and Deckchairs" (Gould and Shearer), 125n20
Bohm, David J., 19
Böhme, Jakob, 166
Bohr, Erik, 43
Bohr, Margrethe, 36–37
Bohr, Niels, 55, 87, 106; Einstein and, 7, 19, 28, 31, 35, 114; Copenhagen legend and, 36–47; Jungk and, 38–41, 43; natural unity and, 54; atomic structure and, 133; scientific context and, 133–134; thematic analysis and, 138, 141, 145, 147
Bolyai, János, 119, 123
Bonolis, Luisa, 63
Boring, Edwin G., 70, 132–133
Bork, Alfred M., 125n20
Born, Max, 19, 27–28, 87, 106
Bose, Satyendranath, 19
Bothe, Walther, 40n6
Bowen, Robert, 174n19
Bragg, William H., 50
Brain, Robert, 170n11
"Brain and the Computing Machine, The" (Weiner), 69
Branscomb, Lewis, 153n4, 156n6, 158
Brauer, Jerald C., 114
Brave New World (Huxley), 63
Breton, André, 127
Bricmont, Jean, 163n6
Bride Stripped Bare by Her Bachelors, Even (Large Glass) (Duchamp), 127–128
Bridgman, P. W.: Mach and, 67; International Congress for the Unity of Science and, 69–70; description of, 72–73; Harvard and, 73; Nobel Prize and, 73, 135; publications of, 73–76; methods of, 73–77; Skinner and, 77–80
Brighter Than a Thousand Suns: A Personal History of the Atomic Scientists (Jungk), 38–41
Broch, Hermann, 134

Brooks, Harvey, 153
Brown, George E., Jr., 152
Brownian motion, 7, 22–23
Bruer, J. T., 193n5
Brush, S. G., 203n4
Buber, Martin, 105
Buck, Barbara, 64
Buddha, 9
Burtt, E. A., 165
Bush, Vannevar, 34
Byron, George Gordon, 166

Cabanne, Pierre, 127
Cannon, W. B., 68
Capon, Augusto, 63
Carnap, Rudolf, 65, 69
Carr, Peter, 197–198
Carrara, Nello, 50
Carter, Jimmy, 158–160
Caruccio, A., 61n3
Carus, Paul, 10
Cassidy, David, 40n6, 45, 203n4
Castelnuovo, Guido, 51, 53
Catholicism, 3
Causality, 28, 32, 106, 110, 142. *See also* Philosophy
Cavendish Laboratory, 57
CERN (Conseil Européen pour la Recherche Nucléaire), 90
Cerullo, John J., 66n1, 70–71
Chadwick, James, 55
Chamberlain, Houston Stewart, 177
"Chance and Necessity in Fermi's Discovery of the Properties of Slow Neutrons" (De Gregorio), 61n3
Chandrasekhar, Subrahmanyan, 54, 58, 60n2, 61
"Changing Structure of the U.S. Research System, The" (Brooks), 153n5
Chaos theory, 115n1
Church, F. Forrester, 114
Cloud chamber, 55–56
Cold War, 8, 152
Cole, Jonathan R., 186, 193n5
Cole, Stephen, 186
Coleridge, Samuel Taylor, 37
Collected Papers (Fermi), 51, 54, 58, 61, 64

Collected Papers (Einstein), 20, 23–25
Columbia University, 34, 57, 62, 81, 86, 88, 198, 201
Commonweal journal, 108
Comparative Physiology of the Brain and Comparative Psychology (Loeb), 66
Conant, James Bryant, 137, 195, 198–199
"Concept of the Reflex in the Description of Behavior, The" (Skinner), 67
Concepts and Categories (Berlin), 144
Condon, Edward U., 87–88
Conservation of energy, 28
Conservation of momentum, 28
Contemporary Civilization program, 198
Continuum, 142, 146
Copenhagen (Frayn), 36–43, 46
Copernican theory, 84–85, 165
Corbino, Orso Mario, 51–53
Core Curriculum, 199–201
Cornell University, 85
Cosmology, 107
Cosmological constant, 14
Creationism, viii
Creativity, 188, 191
"Crisis of the Sciences as Expression of the Radical Life-Crisis of European Humanity, The" (Husserl), 196
"Critique of Psychoanalytic Concepts and Theories" (Skinner), 72
Critique of Pure Reason (Kant), 149, 167
Crombie, A. C., 149
Crooked Timber of Humanity, The: Chapters in the History of Ideas (Berlin), 2, 162n1, 163n1, 164n7
Crout, William, 95
Crozier, W. J., 66, 70
Crystals, 50, 54, 86
Cubists, 124
Cummings, E. E., 134
Curie, Marie, 19, 190
Curie-Joliot, Irene, 56–57
Cybernetics, 69
Cyclotron, 55–56
Czechoslovakia, 33

Dada group, 118
Daedalus journal, 95, 205n6, 211

D'Agostino, Oscar, 57
d'Alembert, Jean le Rond, 65
Dalton, John, 207
Darrow, K. K., 146–147
Davisson, Clinton S., 55
Davos, Profil eines Phanomens (Halter), 114
Davos meeting, 102–106, 109
De Benedetti, Sergio, 55
de Broglie, Maurice, 19, 50
Debye, Peter J. W., 19, 50
Decline of the West, The (Spengler), 172
Defending Science—Within Reason: Between Scientism and Cynicism (Haack), 163n5
De Gregorio, Alberto, 61
Deichmann, Ute, 173n17
Democritus, 11, 142
de Pawlowski, Gaston, 124
Depression, 98
Descartes, René, 206
de Sitter, Willem, 7–8, 107
Determinism, 105–106
Deutsch, Karl, 70n6, 139
Dewey, John, 195
Diaghilev, 133
Dictionary of Scientific Biography, 119n5
Dieudonné, Jean, 118
Dirac, P. A. M., 55, 87
Discontinuum, 146–147
"Discussions with Einstein on Epistemological Problems in Atomic Physics" (Bohr), 31n6
Disease, 159
Donne, John, 165, 196–197
Dostoyevsky, 124
Dreier, Katherine S., 128
Dresselhaus, Mildred, 192
Drude, Paul, 23
Du Bois-Reymond, Emil, 100, 168
DuBridge, Lee, 86
Duchamp, Marcel, 117; Dada group and, 118; French anarchism and, 118; higher-dimensional geometry and, 125–129; Jouffret and, 129–130
Duchamp in Context (Henderson), 118n4, 123

Du Châtelet, Émilie, 194
Dukas, Helen, 16–21, 24–25
Dynamics of Faith (Tillich), 106
Dyson, Freeman, 19

Eddington, Arthur Stanley, 19, 50
Edizioni Scientifiche Società Italiana di Fisica (Bernardini and Bonolis), 63
Edsall, John, 69
Education: specialization and, 195–197; general policies for, 197–199; Core Curriculum and, 199–201; College Cafeteria metaphor of, 200–201; consensus and, 201; new approaches for, 201–208; overlapping subjects and, 202, 205–208; cultural issues and, 202–208; course presentation and, 203–205. *See also* Science
Ego and His Own, The (Stirner), 118
Ehrenfest, Paul, 8, 19, 31, 52
Ehrenhaft, Felix, 143
Einaudi, Renato, 55
Einstein, Albert, 68, 75, 100; Bible and, 3–4; escapism and, 3–4; "Autobiographical Notes" of, 3–4, 6–7, 18n1, 21, 23, 109; religion and, 3–4, 9–15, 84, 108–110; Third Paradise of, 3, 13, 15; on science, 4; world view of, 4–9, 14, 107; thought experiments and, 6; despair and, 6–7; relativity and, 6–8, 14 (*see also* Relativity theory); generalization and, 7–8; nationalism and, 8; politics and, 8–9; God and, 9–13, 31, 108–111, 113; Spinoza and, 10–15, 109; cosmological constant and, 14; Dukas and, 16–20, 24–25; Roosevelt letters and, 19, 33–35; anti-Semitism and, 21, 28, 108; Michelson-Morley experiment and, 22; Brownian motion and, 22–23; Maxwell's theory and, 23; Marić and, 23–24; on the ether, 24; positivism and, 26; Heisenberg and, 26–35; Nazis and, 28; uncertainty principle and, 31; atomic weapons and, 33–34; FBI and, 34; self-study and, 50; Weyl and, 51; Nature's unity and, 54; tacit knowledge and, 60; Rabi and, 82;

Tillich and, 99, 102–114; Davos meeting and, 102–106, 109; determinism and, 105–106; causality and, 106, 110; Copenhagen interpretation and, 106; Maxwellian Program of, 109; Bohr and, 114; Euclid and, 119; Frank and, 137; presuppositions and, 137–138; thematic analysis and, 137–138, 141–146, 149–150; continuum and, 138; Mach and, 138; Reichenbach and, 138–139
Einstein, B to Z (Stachel), 22
Einstein, Elsa, 13–14, 19
Einstein, History, and Other Passions (Holton), 151
Einstein, Maja, 3–4
Einstein and Religion (Jammer), 13
Einstein Archive, 17–21, 23–25, 29
"Einstein's Germany" (Stern), 196n1
Eisenhower, Dwight D., 92
Electromagnetism, 23–24, 29, 86, 171
Elementary Treatise on the Geometry of Four Dimensions: An Introduction to the Geometry of n-Dimensions (Jouffret), 123
Elements (Euclid), 119
Elkana, Yehuda, 196n1
Emotional life (*Gefühlsleben*), 5
Empiricism, 140–141
Encounters with Einstein (Heisenberg), 35
Encyclopedia of Library and Information Science, 148n7
Encyclopedic solution, 197
End of Science, The: Facing the Limits of Knowledge in the Twilight of the Scientific Age (Horgan), 163
Energy, 8
Enlightenment, 168
Enrico Fermi, Physicist (Segrè), 59, 64
"Enrico Fermi a la scoperta degli effetti delle sostanze idrogenate sulla radioattività indotta dei neutroni" (De Gregorio), 61n3
"Enrico Fermi at Pisa" (Gambassi), 64
Enriques, Federico, 51
Equivalence principle, 6
Eranos-Jahrbuch, 135

Erasmus, 207
Erforschtes und Erlebtes (von Harnack), 172n15
Erikson, Erik, 49, 199
Ernst, Max, 124
"Ernst-Mach-Verein" (Vienna Circle), 65
Escapism, 3–4
Essential Tillich, The (Church), 114
Ether, 21, 24
"Ether Problem, the Mechanistic Worldview, and the Origins of the Theory of Relativity, The" (Hirosige), 21n3
Ethics (Spinoza), 12–14, 109
Euclidean geometry, 4, 119–121
Evolution, viii
Existentialism, 102
"Experience and the Law of Causality" (Frank), 150n9

Facing Up: Science and Its Cultural Adversaries (Weinberg), 152n2, 163n6
"False Dichotomy, The: Scientific Creativity and Utility" (Branscomb), 153n4, 158n8
Fano, Ugo, 55
Faraday, Michael, 6, 8, 50, 146, 171
Farm Hall, 39–40, 46
Fascism, 51, 62
Fasetti, Franco, 51
Fashionable Nonsense: Postmodern Philosophers' Abuse of Science (Sokal and Bricmont), 163n6
Faulkner, William, 134
Feenberg, Eugene, 55
Fellows of the Royal Society, 22
Fermi, Enrico, 34, 121, 188–189; moderator effect and, 48–49; background of, 49–50; organizing ability of, 50; publications of, 50–52, 56–57, 59–60; relativity theory and, 50–51; Corbino and, 51–55; Roman School and, 52–53, 55, 57–58, 62–63; teaching methods of, 53; Nature's unity and, 54; study approaches of, 54, 62; collaborations of, 54–55; artificial radioactivity and, 54–62; quantum theory

Fermi, Enrico (continued)
and, 55; paraffin experiment of, 59–61; intuition of, 59–62; Nobel Prize and, 62
Fermi, Laura, 58n1, 63–64
Fermi coordinates, 51
Fermi e la fisica moderna (Pontecorvo), 55, 64
"Festschrift for I. I. Rabi, A" (Motz), 83n3
Fichte, Johann Gottlieb, 101, 169
Fingerspitzengefühl (fingertip feeling), 60, 75
Fisher, Michael, ix
Fluctuation phenomena, 23
Föppl, Ludwig, 23
Formalism, 74
Foundational Debate, The (Köhler, De Pauli-Schimanovich, and Stadler), 69n5, 136n1
Foundations of Scientific Method: The Nineteenth Century (Giere and Westfall), 169n10
Fourth Dimension and Non-Euclidean Geometry in Modern Art, The (Henderson), 123, 129nn22–25
"Fourth Dimension in Nineteenth-Century Physics, The" (Bork), 125n20
Frank, Philipp, 17, 69, 70n6, 72n9, 75n13, 136–137, 150n9, 195
Franklin, Benjamin, 92
Frayn, Michael, 36–43, 46
Freeman, Edward M., 71
Freud, Sigmund, 5, 19, 85, 102
Friedmann, Alexander, 14
Frisch, Max, 36
Frisch, Otto, 48
"From Hamilton College to Walden Two: An Inquiry into B. F. Skinner's Early Social Philosophy" (Wiklander), 66n1
From the Closed World to the Infinite Universe (Koyré), 166n9
Fubini, Eugenio, 55
fuchsian functions, 118, 120–121
"Fundamental Concepts of Physics and Their Most Recent Changes" (Einstein), 114
"Fundamental Concepts of Physics in Its Development, The" (Einstein), 104

"Fundamental Ideas and Methods of Relativity" (Einstein), 6
Future of Religion, The (Tillich), 101, 114

Galileo, 21, 119, 142, 165–166, 196, 206
Galison, Peter, 139n5
Gambassi, Andrea, 64
Gamow, George, 14
Gandhi, 19
Garbasso, Antonio, 55
Gauss, Carl Friedrich, 118–119, 121
Geiger counters, 55–57
Gender Differences in Science Careers: The Project Access Study (Sonnert), 182n2
"General Education in a Free Society," 198–199
"General Relativity Theory" (Einstein), 6
Genesis, book of, 84
George, Stefan, 101
Germaine de Stael, Madame, 194
Germany, 28, 32, 57, 63; World Ice Theory and, 33, 174–175, 177; Copenhagen legend and, 36–47; Rabi and, 87–88; poetry and, 101; religion and, 101; Spengler and, 172–173; Nazi science and, 172–180
Germer, Lester, 55
Gespräche mit Hitler (Rauschning), 23, 176n21, 177n22
Gestalt switch, 146
Giere, R. N., 169n10
Glatzer, Nahum, 105
Gleizes, Albert, 124
Global warming, viii
God, 165–167, 169; Einstein and, 9–13, 31, 108–111, 113; causality and, 32, 110; Heisenberg and, 32; Rabi and, 84–85, 92; Tillich on, 96 (*See also* Tillich, Paul); physics and, 97; Divine Will and, 110
Gödel, Kurt, 134, 185
Gode-von Aesch, Alexander, 164n8
Goebbels, Josef, 175–176
Goethe, 12, 101, 107, 168
Goldstein, Herbert S., 9–10
Gombrich, E. H., 132
Goroff, Daniel L., 115n1
Gould, Stephen Jay, 125n20

Gravity, 6, 107
Great Books program, 198
Grossmann, Marcel, 7
Groth, Wilhelm, 33
Grundkraft, 100
Gustin, Bernard H., 10

Haack, Susan, 163n5
Habicht, Conrad, 108
Habilitationsvortrag, 139
Hadamard, Jacques, 8, 10, 120nn7,9
Haeckel, Ernst, 10
Hahn, Hans, 65
Hahn, Otto, 40, 48–49
Halter, Ernst, 114
Hamilton, George, 127
Hamilton, Richard, 127
Hamilton College, 65
Hamlet (Shakespeare), 101
Handbuch (Pauli), 27
Harrington, Anne, 173
Harteck, Paul, 33
Harvard University, 20, 65, 182; Skinner and, 66–69, 71–72; Society of Fellows and, 68; William James Lectures and, 70; Bridgman and, 73; Morris Loeb lectures and, 83; Divinity School and, 95–96; Nobel Prizes of, 182; "General Education in a Free Society" and, 198–199
Harvard University Press, 135
Hawking, Stephen, 166
H. C. Ørsted and the Romantic Legacy (Brain and Knudsen), 170n11
H. C. Ørsted's Scientific Papers (Meyer), 170n11, 171n14
Hebrew University, 25
Hegel, G. W. F., 97, 101, 132, 168
Heidegger, Martin, 97
Heilbron, J. L., 148n7
Heisenberg, Werner, 7, 50, 55, 87, 141–142; education of, 27; quantum theory and, 27–33, 35; meets Einstein, 28–29; relativity theory and, 29; observability and, 29–31; neo-Platonism and, 30; uncertainty principle and, 30–31; Nobel Prize and, 31; Solvay Congress and, 31–32; anti-Semitism and, 32–34; God and, 32; attacks on Einstein, 33–35; *Copenhagen* play and, 36–43; Copenhagen legend and, 36–47; *Uranverein* and, 38, 40–42, 46; Jungk and, 38–41; atomic weapons and, 39–41; Pusey and, 96; Copenhagen interpretation and, 106; discontinuum and, 146; thematic analysis and, 146–147
Helgoland, 29
Helmholtz, Hermann von, 23–24, 89, 100, 121, 168, 194
Henderson, Lawrence J., 66, 69
Henderson, Linda Dalrymple, 118n4, 123n19, 129
Henri Poincaré (Lebon), 115n1
Heraclitus, 101, 142
Herder, Johann Gottfried, 164, 176
Hertz, Heinrich R., 23–24, 100, 146
Hilbert, David, 5, 19
Himmler, Heinrich, 32–33, 175
Hirosige, T., 21n3
"Historical Perspective on Copenhagen, A" (Cassidy), 40n6
History of Quantum Mechanics Project, 26
History of the Theories of Ether and Electricity, A (Whittaker), 21–22
History of Twentieth-Century Physics (Weiner), 64
Hitler, Adolf: Nazi science and, 172–180; intellectual education and, 176–177
Hitler's Uranium Club: The Secret Recordings at Farm Hall (Bernstein), 39n4, 40n6
Hoagland, Hudson, 69
Holbach, Paul H. T., 180
Hölderlin, Friedrich, 101
Holism, 100, 112
Holton, Gerald: Dukas and, 16–21; research methods of, 16–25; meets Heisenberg, 26, 31; International Congress for the Unity of Science and, 69–70; Tillich and, 95–96; thematic analysis and, 135–151; Einstein's concepts, 138–139

Homme intrigue par le vol d'une mouche non-euclidienne, L' (Ernst), 124
Hoover, J. Edgar, 34
Hörbiger, Hanns, 174
Hörbigers Glacial-Kosmogonie (Hörbiger), 174
Horgan, John, 163n6
House Built on Sand, A: Exposing Postmodernist Myths about Science (Koertge), 163n6
Hubble, Edwin, 14
Humboldt, Alexander von, 194
Husserl, Edmund, 196–197
Hutchins, Robert Maynard, 195
Huxley, Aldous, 63
Huxley, Leonard, 182n1
Huxley, Thomas H., 181–182
Hydrogen bomb, 90

Idealism, 102
Ideas and Opinions (Einstein), 10
Indiana University, 69
Infeld, Leopold, 19
In Search of Cultural History (Gombrich), 132n27
Institute for Scientific Information (ISI), 187
Institute for Advanced Study, 17, 19–20, 137–138
Instrumentalism, 27–28, 135
International Congress for the Unity of Science, 69
International Encyclopedia of Unified Science (Neurath), 198
"Invention mathématique, L'" (Poincaré), 120
Ionian Fallacy, 144–145
Iron Curtain, 90
Isotopes, 56
Israel Academy of Sciences and Humanities, 17
Italian Physicists and Their Institutions (Buck), 64
Italy, 48–51; Roman School and, 52–53, 55, 57–58, 62–63
Ivory Bridges: Connecting Science and Society (Sonnert), 159n9

Jackson, Andrew D., 170n11
Jacobi, K. G. J., 86
Jakobson, Roman, 70, 139n5
James, William, 134
Jammer, Max, 13–14
Jarry, Alfred, 124
Jefferson, Thomas, 197–198
Jeffersonian research, 155–161, 184, 194
Jelved, Karen, 170n11
Jesus Christ, 9
Joliot, Frédéric, 56–57
Josephson, Paul R., 173n17
Jouffret, E., 123, 125, 129–130
Judaism, 3, 9–10, 84
Jungk, Robert, 38–44

Kaiser, Rudolf, 105
Kandinsky, Wassily, 133
Kant, Immanuel, 100–101, 134, 181, 206; Fermi and, 60; thematic analysis and, 140, 148–150; postmodernism and, 167–172
"Kant, 'Naturphilosophie' and Scientific Method" (Williams), 169n10
Kastorp, Hans, 104
Kayser, Rudolf, 12
Keats, John, 37, 47, 166
Kemble, Edwin C., 70n6
Kent, Allen, 148n7
Kepler, Johannes, 50, 89, 136–137, 164–165, 171, 206
Kimball, Robert C., 114
Kistiakowsky, George, 159n10
Klinger, F. M., 169
Knowledge and Error (Mach), 66
Knudsen, Ole, 170n11
Koch, Caesar, 23
Koertge, Noretta, 163n6
Köhler, E., 69n5
Koyré, Alexandre, 165, 166n9
Kragh, Helge, 139n5
Kramers, Hendrik A., 28
Krieck, Ernst, 178–179
Krisis der europäischen Wissenschaften und die phaenomenologische Philosophie, Die (Husserl), 196
Kronig, Ralph, 86

Kuhn, Thomas S., 24
Kulturbund, 196
Kuznetsov, Boris G., 17

Langevin, Paul, 121
Laporte, Otto, 27–29
Latour, Bruno, 152n2
Lawrence, E. O., 55–56
Laws, Joan, ix
Lebon, Ernest, 115n1
Lederman, Leon, 88
Lee, Paul, 96
Lehre von Sein, Die (Hegel), 97
Leibnitz, Wilhelm Gottfried, 134
Lenard, Philipp, 28
Leontief, Wassily, 69
Levi, Primo, 61
Levi-Civita, Tullio, 51, 53
Lévy-Bruhl, Lucien, 104
Lewin, Kurt, 69, 149
Lewis and Clark expedition, 156
Life and Letters of Thomas Henry Huxley, The (Huxley), 182n1
"Life and Work of Walter Cannon, The" (Edsall), 69
Light, 7, 30, 109
"Living Philosophers" series, 3
Living Philosophies (Einstein), 114
Lobachevsky, Nicolai Ivanovich, 119, 123
Loeb, Jacques, 5, 66, 68
Logan, Jonothan, 41n7
Logic of Modern Physics (Bridgman), 67, 76, 77n16, 135–136
London, Fritz, 55
Long, S. J., 187n4
Lorentz, Hendrik Antoon, 19, 21–22
Losee, John, 139n5

Mach, Ernst, 5, 26, 66–69, 138, 142–143, 150
MacLeish, Archibald, 134
Magic Mountain, The (Mann), 104
Magnetic resonance imaging, 202
Magneto-optics, 51
Majorana, Ettore, 53
Malevich, Kazimir, 124
Mandarins, 70–71

Manifesto of a Passionate Moderate (Haack), 163n5
Mann, Thomas, 19, 104, 134
Mannheim, Karl, 70
Marcel Duchamp (Paz), 127n21
Margenau, Hans, 70n6
Marić, Mileva, 23–24
Marshall, John, 53
Marxism, 85, 102
Mass, 8
Materials science, 73
Mathematics, 50–51, 86–87, 102, 194; $1/v$ law, 59; Poincaré and, 115 (*See also* Poincaré, Henri); fuchsian functions and, 118, 120–121; n-dimensionality and, 119, 123–131; Euclidean geometry and, 119–121; non-Euclidean geometry and, 121–124; motor space and, 122; Jouffret and, 123, 125, 129–130; Duchamp and, 125–129; art and, 125–134; good science and, 183
Maud Committee, 34
Maxwell, James Clerk, 23, 30, 86, 100, 144, 146, 171
"Measures of Sex Differences in Scientific Productivity" (Long), 187n4
Mechanics (Mach), 67
Medicus, Fritz, 105
Mein Kampf (Hitler), 174, 176
Meitner, Lise, 19, 48, 54
Merton, Robert K., 10, 139n5
Metaphysical Foundations of Natural Science (Kant), 100, 167–168
Metzger, Helene, 165
Metzinger, Jean, 124
Meyer, Kristine, 170n11, 171n14
Michelson, A. A., 142
Michelson-Morley experiment, 22
Millikan, R. A., 54, 142
Mind Always in Motion, A: The Autobiography of Emilio Segrè (Segrè), 64
Mind in the Modern World (Trilling), 195
Minkowski, Hermann, 19
MIT Radiation Laboratory, 81
Mittag-Leffler, G., 115n1
Moderator effect, 48–49

Modern Theme, The (Ortega y Gasset), 132
Molecular beams, 88–89
Møller, Christian, 44
Mongardini, C., 193n5
Monist, The (Carus), 10
Moral issues, 36–47
Morgenstern, Oscar, 70
Morison, Robert, 18
Morris, Charles, 69
Morris Loeb lectures, 83
Moseley, Henry G. J., 50
Moses, 9
Motive des Forschens (Einstein), 60
Motz, Lloyd, 83n3
Mozart, Wolfgang Amadeus, 121
Murray, Henry A., 70
Mussolini, Benito, 51
My Search for Absolutes (Tillich), 111–112, 114
Myth of the Twentieth Century, The (Rosenberg), 176

Nagel, Brigitte, 174n19
Nagel, Ernest, 70n6
Nathan, Otto, 21, 24
National Academy of Sciences, 83n3, 203
National Institutes of Health (NIH), 153–154, 157
Nationalism, 8, 71
National Research Council, 56–57, 62, 68
National Science Education Standards, 203
National Science Foundation, 82
Natural philosophers. *See* Philosophy
Natural Science in German Romanticism (Gode-von Aesch), 164n8
Nature, 55
"Nature Revealed: The Joys and Dangers of Experimental Physics" (Rabi), 83n3
n-dimensional manifolds, 119, 123–131
Nedermeyer, Seth, 55
Nemorov, Howard, 98, 114
Neo-Platonism, 30, 134
Neurath, Otto, 65, 75n12, 198
Neutrino hypothesis, 55
Neutrons, 48–49, 54
New Methods of Celestial Mechanics (Goroff), 115n1

Newton, Isaac, 21, 134, 167–168, 202–203, 206–207
Newtonian research, 154–156
"New Vistas for Intelligence" (Bridgman), 76n14
New York Review of Books, The, 41n9
New York Times, 37n1, 46
Niels Bohr Archives, 24, 43
Nietzsche, 101
Nigo, M., 61n3
Nisbet, Robert, 139n5
Nobel Prize, 56, 62, 73, 89, 135, 182
Non-Euclidean geometry, 121–124, 129
Notes and Projects for the Large Glass (Duchamp), 126
Novalis, 101, 168
Nuclear magnetic resonance (NMR), 161
Nuclear science: weapons research and, 33–35, 40–41; *Uranverein* and, 38, 40; moderator effect and, 48–49

"Objectivity of Science, The: A Crucial Problem" (Krieck), 178
Occhialini, Giuseppe, 55
Occultism, 169
O'Connell, Cardinal, 9, 108
Oersted, Hans Christian, 60–61, 100, 144, 170–171, 181, 183
Oken, Lorenz, 169
"On Einstein's Distrust of the Electromagnetic Theory: The Origin of the Lightvelocity Postulate" (Abiko), 18n1
"On the Quantum Theoretical Reinterpretation of Kinematic and Mechanical Relations" (Heisenberg), 29
"On the Validation of Scientific Theories" (conference), 71–72
Open Court, The (Carus), 10
Opera di R. K. Merton e la Sociologia contemporanea, L' (Mongardini and Tabboni), 193n5
"Operational Basis of Psychology, The" (Stevens), 77n17
Operationalism, 27–28, 74
Oppenheimer, J. Robert, 37, 81, 88, 90, 142, 205

Opticks (Newton), 167
Organism as a Whole, The (Loeb), 66
Origins of Postmodernity, The (Anderson), 163n6
Ortega y Gasset, José, 132
Ostwald, Wilhelm, 10, 49, 121, 150, 166, 182–183
Ouspensky, P. D., 124
Outer Circle, The: Women in the Scientific Community (Zuckerman, Cole, and Bruer), 193n5
Oxford Companion to the History of Modern Science, The, 148n7

Pais, Abraham, 17, 41n7
Paradigms, 148–149
Parmenides, 101, 142
Parsons, Talcott, 69, 70n6
Pasteur's Quadrant: Basic Science and Technological Innovation (Stokes), 157–158
Pauli, Wolfgang, 19, 27, 50, 55, 87
Paul Tillich's Philosophy of Culture, Science, and Religion (Adams), 114
Paz, Octavio, 117, 127
Peierls, Rudolf E., 40n6, 42n11, 55
Pelseneer, Jean, 23
Perisco, Enrico, 55, 58, 60n2
Perrin, Jean, 133
Pestalozzi, Johann, 195
Phenomenology, 102
Philosophy, 3, 17, 60; positivism, 5, 26, 66, 132–133, 140–141; Spinoza and, 10–15; probabilism, 13; rationalism, 26; instrumentalism, 27–28, 135; neo-Platonism, 30, 134; tacit knowledge, 60–61; Bridgman and, 72–80; operationalism, 74; Marxism, 85; Tillich and, 95–114; cultural influences and, 99–100; Holism, 100; Kant and, 100; natural, 100–101; synthesis, 100–101, 113, 124–125, 142, 144, 181; determinism, 105–106; French anarchism, 118; Hegelian intuition, 132; *Zeitgeist* and, 132–133; pragmatism, 135; thematic analysis and, 135–151; empiricism, 140–141; incommensurability and, 142; dyads and, 142–143, 147–148; paradigm concept and, 148–149; thought categories and, 148–150; *a priori* reasoning and, 149–150; romanticism, 162–164, 166, 167–180; postmodernism, 162–180; mind-body dualism, 165–166; scientific specialization and, 195–197; Jefferson and, 198; education policies and, 205–208. *See also* God
Physical Review, 88
Physics: moderator effect and, 48–49; artificial radioactivity and, 54–62; materials science and, 73; Project Physics Course and, 82; God and, 97. *See also* Science
Physics, the Human Adventure: From Copernicus to Einstein and Beyond (Holton and Brush), 201n3
Physics in Perspective, 64
Physics Today, 64, 211
Piaget, Jean, 104
Picasso, Pablo, 117, 129, 131
Placzek, George, 55
Planck, Max, 5, 108, 154–155; thematic analysis and, 138, 142, 144–145, 150
Plato, 73, 125
Platonic bodies, 137
Pleijel, Hans, 62
Plummer Cobb, Jewell, 193
Poincaré, Henri, 22, 60, 67; positions held by, 115; accomplishments of, 115–116; fuchsian functions and, 118, 120–121; ignorance of mathematical literature, 118–119; Euclidean geometry and, 119–121; conservatism of, 120–122; methods of, 120–122; unconscious ideas and, 120–122; non-Euclidean geometry and, 121–124; n-dimensionality and, 123–124; visualization and, 124–125
Polanyi, Michael, 60
Politics of Cultural Despair: A Study in the Rise of the Germanic Ideology (Stern), 173n17
Pontecorvo, Bruno, 55, 58, 64

Pope Urban VIII, 166
Popper, Karl, 9, 134
Portmann, Adolf, 139n4
Portrait of Wilhelm Uhde (Picasso), 129, 131
"Portraits of Fermi" (Bethe), 64
Positivism, 5, 26, 66, 132–133, 140–141
Positrons, 56
Postmodernism, viii; romanticism and, 162–164; charges against, 164–167; Kant and, 167–172; Nazi science and, 172–180
Pound, Ezra, 134
Powers, Thomas, 41–42, 45–46
Pragmatism, 135
"Present State of Operationalism, The" (Bridgman), 74n11
Press, Frank, 88–89, 158–160
Princet, Maurice, 123
Princeton University, 16–17, 19–21, 137–138
Principia (Newton), 154, 167
Principles of Physical Optics, The (Mach), 66
"Prinzipien der Forschung" (Einstein), 4–5
Probabilism, 13
Proceedings of the International Conference: Enrico Fermi and the Universe of Physics, 64
Project Access, 182–188
Project Physics Course, 82, 90
"Prologue, Perspectives, and Prospects of Behaviorology" (Vargas), 66n2
Proust, Marcel, 124
"Psychoanalysis and the Theory of Social Systems" (Parsons), 69
Psychology of Invention in the Mathematical Field, The (Hadamard), 8, 10, 120n7, 121n9
Pugwash movement, 8
Pusey, Nathan, 96

Quantum Physics Oral History Project, 18, 83, 84n4, 88n17, 89n19
Quantum theory, 7, 22, 35, 142; radiation structure and, 23; History of Quantum Mechanics Project and, 26; Heisenberg and, 27–33; complexity of, 29; observability and, 29–31; causality and, 32; Fermi and, 50, 55; beta decay and, 55; Rabi and, 83–84, 86–87; Copenhagen interpretation of, 106
Questor Hero, The: Myth as Universal Symbol in the Works of Thomas Mann (Nemorov), 98, 114
Quine, W. V., 65, 69–70, 75n13, 136

Rabi, I. I., 57, 97; publications of, 81; vision of, 81; Project Physics Course and, 82; education and, 82–89; Quantum Physics Oral History Project and, 83; background of, 83–84; quantum theory and, 83–84, 86–87; Bible and, 84–85; Copernican theory and, 84–85; God and, 84–85, 92; mindset of, 84–85; self-study and, 85–87; study groups and, 86; historic consequence and, 87; American laboratories and, 87–88; Germany and, 87–88; work ethics of, 88; molecular beam research and, 88–89; Nobel Prize and, 89; as science warrior, 89–92; politics and, 90–92; anti-Galileans and, 91
Rabi, Scientist and Citizen (Rigden), 83
"Rabi: The Modern Age" (Bernstein), 84n5
Racah, Giulio, 55
Radiation, 23–24, 30; atomic theory and, 54–62
Ramsey, Norman, 83n3, 88
Rasetti, Franco, 50, 52–54, 57–58
Rathenau, Walther, 108
Rationalism, 26
Raum, Zeit, Materie (Weyl), 27, 50–51
Rauschning, Hermann, 23, 176n21, 177n22
Red Book, 198–199
Re-enchanted Science: Holism in German Culture from Wilhelm II to Hitler (Harrington), 173n17
Reflections of a Physicist (Bridgman), 75, 76n14

Reforming of General Education, The (Bell), 201n3
Reichenbach, Hans, 135, 138–139, 198
Reid, Herbert, 139, 152
Relativity theory, 139; general theory of, 6–8, 14, 27, 107–108, 134; cosmological constant and, 14; Einstein's first book of, 21; special theory of, 21–24, 27, 107; Lorentz transformations and, 22; Michelson-Morley experiment and, 22; Brownian motion and, 22–23; Heisenberg and, 26–27, 29; Pauli and, 27; instrumentalism and, 27–28; quantum theory and, 32–33; Fermi and, 50–51; operationalism and, 74; cosmology and, 107; public issues over, 108–109; semantic issues and, 108–109; Tillich and, 111–112
"Religious Symbol, The" (Tillich), 95
Religion: Einstein and, 3–4, 9–15, 84, 108–110; cosmic, 84; Rabi and, 84–85, 92; Tillich and, 95–114 (*See also* Tillich, Paul); German Socialism and, 101; Schelling and, 101; Personal God and, 110–111
"Religion, Art, and Science" (Tillich), 113
"Religion and Culture" (Tillich), 104
"Religion and Science" (Einstein), 109
Religion of Science, The (Carus), 10
Religiöse Reden (Tillich), 114
"Religious Knowledge, The" (Tillich), 104
"Religious Situation, The" (Tillich), 111
"Resistance of 72 Elements, Alloys, and Compounds to 100,000 Kilograms per Square Centimeter, The" (Bridgman), 73
Ricci-Curbastro, Gregorio, 51
Ricerca Scientifica, 56–57, 59
Richards, I. A., 69–70
Richter, Hans, 139
Riemann, Bernhard, 119, 139
Riesman, David, 199
Rigden, John, 12–14, 16, 83, 84n6, 85nn7–9, 86nn10,12–14, 87nn15,16, 90n21, 91nn23–27, 92nn28–33

Righi, Augusto, 51
Rilke, Rainer Maria, 101
Rite of Spring (Stravinsky), 133
Ritter, Johann Wilhelm, 168–169
Robertson, Bob, 87–88
Rockefeller Foundation, 18
Rolland, Romain, 12, 19
Roman School, 52–53, 55, 57–58, 62–63
Romanticism, 100, 162–164, 166; Kant and, 167–172; Nazis and, 172–180
Roosevelt, Franklin D., 19, 33–35
Roots of Romanticism, The (Berlin), 163n4
Rosenberg, Alfred, 176
Rosenzweig, Franz, 105
Rossi, Bruno, 55, 58, 60n2
Rückblick (Kandinsky), 133n29
Russell, Bertrand, 19, 66
Rust, Bernhard, 177–178
Rutherford, F. J., 47, 89, 148

St. Francis of Assisi, 11
St. John's College, 198
Samuelson, Paul, 70n6
Sarton, George, 66, 69
Schelling, Friedrich, 101, 111–112, 168
Scherrer, Paul, 50
Schiller, Friedrich, 54, 99, 165–166
Schilpp, Paul A., 3, 31n6, 119n6, 150n8
Schimanovich, W., 69n5
Schlegel, A. W., 168, 170
Schlegel, Friedrich, 12, 168, 170
Schleiermacher, Friedrich, 12, 101, 195
Schlick, Moritz, 198
Scholem, Gershom, 139
Schopenhauer, Arthur, 60, 105, 177
Schrödinger, Erwin, 55, 63, 141–142; Einstein and, 19, 31; teaching methods of, 53; Rabi and, 86–87; continuum and, 147; thematic analysis and, 147
Schubert, Franz, 98
Schwarze Korps, Das (SS journal), 32
Schweber, S. S., 139n5
Schweitzer, Albert, 12, 19
Science, vii–viii, 68; Einstein on, 4; despair and, 6–7; generalization and, 7–8; religion and, 9–15, 107–108; as

Science *(continued)*
  devotion, 13–15; observability and, 27–31; causality and, 28, 32, 106, 110; conservation principles and, 28; uncertainty principle and, 30–31; Copenhagen legend and, 36–47; *Uranverein* and, 38, 41–42; moral issues and, 39–41; meaningful events in, 48–49; self-study and, 50; natural unity and, 54; artificial radiation and, 54–62; tacit knowledge and, 60–61; International Congress for the Unity of Science and, 69–70; mandarins and, 70–71; materials science and, 73; American laboratories and, 87–88; Rabi's efforts for, 89–92; as unifying force, 90; as ennobling activity, 90–91; cultural defense of, 91; objective thought and, 91; humanistic approach to, 91–92; quest for ultimate and, 95–114; synthesis and, 100–101, 113, 124–125, 181; as spiritual act, 107–108; art and, 125–134; thematic analysis and, 135–151; Kepler's laws and, 136–137; public, 140; defining, 140–141; incommensurability and, 142; continuum and, 142, 146; dyads and, 142–143; third axis of, 142–143; demeaning of, 152; U.S. Congress and, 152–153, 155; public policy and, 152–161; research modes and, 154–161, 184, 194; Carter-Press initiative and, 158–160; postmodernisms and, 162–180; romanticism and, 167–172; Nazis and, 172–180; good science and, 181–193; Project Access and, 182–188; gender issues and, 183–193; mentorship and, 184; publication rates and, 184–187, 190–191; citation rates and, 187; creativity and, 188, 191; socialization differences and, 188–189; professional conduct and, 189–190; humanities and, 194–208; specialization and, 195–197; worldview and, 195–197; general education and, 197–199; twenty-first century approaches for, 201–205; knowledge networks and, 205–206

Science, Philosophy and Religion: A Symposium, 114
Science, Technology and Society in Seventeenth-Century England (Merton), 10
Science and Anti-Science (Holton), 151, 156n6
Science and Cultural Crisis (Walter), 77n18, 78n20
Science and Hypothesis (Poincaré), 67, 116, 122–123
Science and Method (Poincaré), 60, 67, 121
"Science and Religion" (Einstein), 109–110
"Science and the Humanities" (Rabi), 83
"Science and the Liberating Arts" (Rabi), 91
"Science and Theology: A Discussion with Einstein" (Tillich), 110–111
Science for All Americans (Rutherford and Ahlgren), 148
Science for Society: Cutting-Edge Basic Research in the Service of Public Objectives (Branscomb, Holton, and Sonnert), 156n6
Science Journal Citation Reports, 187
Science of Mechanics (Mach), 66
"Science: Public or Private" (Bridgman), 75
Science: The Center of Culture (Rabi), 81
Scientific Growth (Ben-David), 10
Scientific Imagination, The: Case Studies (Holton), 151
Scientific Philosophy: Origins and Developments (Stadler), 69n5
"Scientific Work of Einstein, The" (Heisenberg), 33
Scientists under Hitler: Politics in the Third Reich (Beyerchen), 173n17
Scott, Grant F., 37n2
Segrè, Emilio, 52–54, 57–59, 60n2, 63–64
Selected Letters of John Keats (Scott), 37n2
Selected Scientific Works of Hans Christian Ørsted (Jelved, Jackson, and Knudsen), 12, 170n11, 171n13
Self-study, 50, 85–87

Shadish, William R., 186
Shakespeare, 101
Shaping of a Behaviorist, The (Skinner), 66
Shapley, Harlow, 70n6, 74
Shaw, Bernard, 19
Shearer, Rhonda Roland, 125n20
Shelley, Percy Bysshe, 166
Simmel, Georg, 188
"Sketch for an Epistemology" (Skinner), 68
Skinner, B. F.: background of, 65–66; "lost years" and, 65–72; Harvard and, 66–69, 71–72; Mach and, 66–69; Ph.D. of, 68; research of, 68–72; as intellectual, 70–71; Bridgman and, 77–80
"Skinner at Harvard: Intellectual or Mandarin?" (Cerullo), 66n1
Slater, John, 28
Smith, L. D., 66n1, 69n4
Snow, C. P., 83
Société de Psychologie, 120
Sokal, Alan, 163n6
Solovine, Maurice, 23
Solvay Conference, 31–32, 61
Sommerfeld, Arnold, 27–28, 50, 87
Sonnert, Gerhard, ix, 152n1, 156n6, 159n9, 180n26, 182n2, 183, 186
Soul in Nature, The (Oersted), 170
Soviet Union, 57
Space, 6–8, 14
Specialization, 195–197
Spectroscopy, 52; artificial radiation and, 54–55, 58, 58–62
Spectroscopy with Coherent Radiation (Ramsey), 83n3
Spengler, Oswald, 172–173
Spinoza, Baruch, 10–15, 60, 109
Spinoza Dictionary, 12
Stachel, John, 19, 22
Stadler, Friedrich, 69n5
Stark, Johannes, 32, 174
State of feeling (*Gefühlszustand*), 5
Stein, Gertrude, 124
Stern, Fritz, 87, 173, 195, 196n1
Stern, Otto, 54
Stevens, S. S., 69, 77n17
Stirner, Max, 118
Stokes, Donald, 157–158

"Strange New Quantum Ethics, A" (Logan), 41n7
Strassmann, Fritz, 48–49
Straus, Ernst, 9, 17
Straus, Raphael, 17
Stravinsky, Igor, 133
Street, J. Curry, 55
Stuart, Mary, 37
Sturm und Drang (Klinger), 169
Styles of Scientific Thinking in the European Tradition (Crombie), 149
Swiss Polytechnic Institute, 8–9, 23
Symmetries, 30
Synthesis, 100–101, 113, 124–125, 142, 144, 181
System der Wissenschaften, nach Gegenständen und Methoden, Das (Tillich), 102, 113
Systematic Theology (Tillich), 96, 101
System of Nature (Holbach), 180
System of the Sciences According to Objects and Methods, The (Tillich), 114
Szilard, Leo, 34

Tabboni, S., 193n5
Tacit knowledge, 60–61
Tagore, R., 19
Teil und das Ganze, Der (Heisenberg), 27n2, 28n3, 29n4, 30n5
Teller, Edward, 55
Thales, 100, 145
Thematic analysis: origins of, 135–139; method of, 140–142; semantics of, 141n6; dyads and, 142–143, 147–148; diversity and, 144–146; aesthetics and, 147; Kant and, 148–150
Thematic Origins of Scientific Thought: Kepler to Einstein (Holton), 20n2, 22, 26n1, 136n2, 142, 148, 204n5
Theology of Culture (Kimball), 114
"Theory of Culture, A" (Tillich), 99
Theory of everything, 8
Theosophists, 124
Thought experiments, 6
Tillich, Hannah, 95
Tillich, Paul, 199; ultimates and, 95, 97–99, 111–112; Nazis and, 96; style of, 96–

Tillich, Paul *(continued)*
  99; infinity and, 98; Einstein and, 99, 102–114; cultural influences and, 99–102; mission of, 99–102; background of, 100–101; Kant and, 100–101; synthesis and, 100–101, 113; poetry and, 101; Schelling and, 101; World War I and, 101–102; publications of, 102, 106, 110–114; Davos meeting and, 102–106; relativity theory and, 111–112
Time of My Life, The: An Autobiography (Quine), 136n1
Tisza, Laszlo, 70n6
Tönnies, Ferdinand, 5
Totalitarian Science and Technology (Josephson), 173n17
Transactions of the New York Academy of Sciences, 83n3
Treatise (Maxwell), 86
Triads, 142–143
Tribute to Professor I. I. Rabi on the Occasion of His Retirement from Columbia University, A, 86n11
Trilling, Lionel, 195–196
Trotzdem: Mein Leben für die Zukunft (Jungk), 41n8
Truman, Harry S, 90
Tucci, P., 61n3
Twentieth-Century Physics, Essays and Recollections: A Selection of Historical Writings by Edoardo Amaldi (Amaldi), 63

Ultimate Concerns, 95
Uncertainty principle, 30–31
Understanding Physics (Cassidy), 201n3
UNESCO, 26
Unification theories, 5–9
Union Theological Seminary in New York, 96
United States, 19, 33–34, 152; American laboratories and, 87–88; Core Curriculum and, 199–201; education policies and, 199–201, 203
Universal Ice (Bowen), 174n19
Universitas, 33n7

University at Göttingen, 27
University of Berlin, 29
University of Chicago, 54
University of Frankfurt, 105
University of Minnesota, 68–69
University of Munich, 27
University of Padua, 48
University of Pisa, 50, 52
University of Rome, 51–52
University of Virginia, 96
Unlocking Our Future: Toward a New National Science Policy, 152n2
Uranium, 33–34, 38–41
Uranium Club (Bernstein), 41n7
Uranverein, 38, 40–42, 46
Urey, Harold, 55
U.S. Department of Agriculture, 159
U.S. Department of Defense, 159
U.S. Department of Energy, 159
U.S. Federal Bureau of Investigation (FBI), 34
U.S. State Department, 159

Valéry, Paul, 19, 124
Validation of Scientific Theories, The (Frank), 72n9, 74n11, 76n15, 133n28
Van Vleck, J. H., 88
Varèse, Edgard, 124
Vargas, E. A., 66
Vienna Circle, 65, 69, 71, 75n12, 136, 198
Virchow, Rudolf, 100, 168
Voice of Memory, The (Levi), 61
Volksgeist, 132
Volterra, Vito, 51
von Harnack, Adolf, 172n15
von Hohenburg, Herwart, 164
von Kleist, Heinrich, 165
von Laue, Max, 7, 23, 28–29, 40, 50
von Liebig, Justus, 170
von Mises, Richard, 69, 136
von Neumann, John, 70
von Schlegel, August Wilhelm, 101
von Weizsäcker, C. F., 39–41, 44, 46

Wagner, Richard, 176
Wald, George, 69, 199
Walden II (Skinner), 71

Walter, Maila L., 77n18, 78n20
Wang, S. C., 86
Watson, John B., 65
Way Things Are, The (Bridgman), 75
Weber, Max, 165
Websites, 40n6, 42n12, 43
Weinberg, Steven, 142, 152n2, 163n6
Weiner, Charles, 64
Weiss, Christian, 169
Weisskopf, Victor, 145
Weizmann, Chaim, 19
Wells, H. G., 124
Welteislehre, Die (Nagel), 174n19
Wentzel, 19
Werther, 98
Westfall, R. S., 169n10
Weyl, Hermann, 27, 50–51
"What Happened at Copenhagen?" (Pais), 41n7
"What Makes a Good Scientist?: Determinants of Peer Evaluation among Biologists" (Sonnert), 186n3
Wheeler, John, 14, 18, 20
Whitehead, Alfred North, 134, 195
Whittaker, Edmund, 21–22
Who Succeeds in Science? The Gender Dimension (Sonnert and Holton), 182n2
Wiener, Norbert, 69
Wiesner, Jerome, 159n10
Wiklander, Nils, 66n1
"Wilhelm von Schutz: A Contribution to the History of the Drama of the Romantic School" (Goebbels), 175–176

William James Lectures, 70
Williams, L. Pearce, 169
Williams, William Carlos, 134
Wimmers, Eckhart, 196–197
Wissenschaftliche Weltauffassung: der Wiener Kreis (Vienna Circle), 65
Wissenschaftslehre (Fichte), 169
Women: Project Access and, 182–188; good science and, 183–193; mentorship and, 184; publication rates of, 184–187, 190–191; evaluation of, 186–187; citation rates of, 187; creativity and, 188, 191; socialization differences and, 188–189; professional conduct and, 189–190; perfectionism and, 190–191
Woodward, W. R., 66n1
World Ice Theory, 33, 174–175, 177
World picture (*Weltanschauung*), 5–9
World War I, 101–102, 161
World War II, 36, 63, 81, 161, 198; *Uranverein* and, 38, 40–42
Wyman, Jeffries, 70

X-rays, 50, 129

Young, Thomas, 142

Zacharias, Jerrold, 88
Zeeman, Pieter, 19, 54
Zeitgeist, 132–133
Zeitschrift für Physik, 29, 86
Zimanski, Mark, 86
Zimmens, Joachim, 104
Zionism, 3–4, 18
Zuckerman, Harriet, 193n5

CENTER
ELIZABETHTOWN COMMUNITY &
TECHNICAL COLLEGE
COLLEGE ST. RD.
TOWN, KY 42701